	DATE DUE		
JAN 02 1996 S			

THE CHEMICAL BOND

STRUCTURE AND DYNAMICS

THE CHEMICAL BOND

STRUCTURE AND DYNAMICS

Edited by

Ahmed Zewail
California Institute of Technology

ACADEMIC PRESS, INC.

Harcourt Brace Jovanovich, Publishers

Boston   San Diego   New York
London   Sydney   Tokyo   Toronto

Copyright © 1992 by Academic Press, Inc.

ACADEMIC PRESS, INC.
1250 Sixth Avenue, San Diego, CA 92101

United Kingdom Edition published by
ACADEMIC PRESS LIMITED
24–28 Oval Road, London NW1 7DX

Library of Congress Cataloging-in-Publication Data

The Chemical bond : structure and dynamics / Ahmed Zewail, [editor]
    p.   cm.
  Includes bibliographical references (p.   ) and index.
  ISBN 0-12-779620-7 (acid-free paper)
  1. Chemical bonds.   I. Zewail, Ahmed H.
QD461.C422   1992
541.2′24—dc20   91-29643
                                                      CIP

Printed in the United States of America

92 93 94 95   9 8 7 6 5 4 3 2 1

To

LINUS PAULING,

one of the greatest scientists of the 20th century

—A. H. Z.

CONTENTS

LIST OF AUTHORS

RICHARD BERNSTEIN
Professor of Chemistry, University of California, Los Angeles, California

FRANCIS CRICK, F.R.S.
J.W. Kieckhefer Distinguished Research Professor for Biological Studies, The Salk Institute, La Jolla, California

DUDLEY R. HERSCHBACH
Baird Professor of Science, Harvard University, Cambridge, Massachusetts

LINUS PAULING
Chairman of the Board and Research Professor, Linus Pauling Institute of Science and Medicine, Palo Alto, California

MAX F. PERUTZ, F.R.S.
Medical Research Council Laboratory of Molecular Biology, Cambridge, United Kingdom

JOHN C. POLANYI, F.R.S.
University Professor and Professor of Chemistry, University of Toronto, Toronto, Ontario, Canada

LORD GEORGE PORTER, O.M. F.R.S.
Chairman of the Centre for Photomolecular Sciences, Imperial College of Science, Technology, and Medicine, London, United Kingdom

ALEXANDER RICH
William Thompson Sedgwick Professor of Biophysics, Massachusetts Institute of Technology, Cambridge, Massachusetts

AHMED ZEWAIL
Linus Pauling Professor of Chemical Physics, California Institute of Technology, Pasadena, California

PREFACE

Understanding the structure and the dynamics of the chemical bond is central to all fields of molecular science. Led by von Laue, the Braggs (father and son), Pauling, and others, from the early days of molecular structure to structural biology, x-ray crystallography has continued to be an important tool for mapping the "static" nature of chemical bonds in substances. More recently, major breakthroughs have been made in determining the structure of "life molecules": DNA, proteins, nucleic acid–protein complexes, viruses, and the photosynthetic reaction center.

The transformation of one substance into another is not a static event, however, and the dynamics and the time scale for the conversion are fundamental to chemical reactivity. Almost a century ago, Arrhenius' famous work provided a macroscopic description of the change in rates of chemical reactions with temperature. Only in the 1940s and 1950s did time resolution, on the milli- to microsecond scale, allow the recording of chemical intermediates. Over the past 30 years, the advent of molecular beams, chemiluminescence, and high-resolution laser spectroscopy has made it possible to study the microscopic behavior of chemical reactions. With ultrashort pulses of light from lasers, time resolution has now reached the time scale of chemical bonding (10^{-14}–10^{-13} second), and, as x-rays introduced the *distance* scale for molecular structure, such laser rays have introduced the *time* scale for the dynamics of the chemical act itself—the transition states between reagents and products.

This volume contains chapters describing the science and the historical developments that have contributed to the progress achieved in the study of both the structure and the dynamics of the chemical bond. In the first part, "Structure," Linus Pauling, Max Perutz, Francis Crick, and Alex Rich give an account of the progress made in different fields with their own historical reminiscences. In his well-known style, Linus Pauling lays the groundwork for this section by discussing the history of x-ray diffraction through the twentieth century—from the days of the structural determination of simple crystals, such as diamond and table salt, by W. H. and W. L. Bragg, to Pauling's own current examination of icosahedral quasicrystals. Max Perutz writes with excitement about the history of his involvement with the hydrogen bond and the significance of this bond in physiology. Alex Rich gives an overview of the three-dimensional structures of proteins and nucleic acids, and the nature of the specific interactions between them—the expression of the genetic information found in nucleic acids. Francis Crick takes a retrospective look at how the understanding of the chemical bond and much of Pauling's work dramatically affected the field of molecular biology. In concluding this section, Pauling offers a personal perspective, "How I Became Interested in the Chemical Bond," that traces his curiosity from a 12-year-old's interest in minerals to his work in x-ray crystallography, which ultimately led to the publication of the classic and monumental book, *The Nature of the Chemical Bond*.

In the second part, "Dynamics," George Porter, John Polanyi, Dudley Herschbach, and the late Dick Bernstein (with the editor) provide overviews of the advances made in developing the molecular description, in time and space, of chemical reactivity, and the impact of recent discoveries. In probing the dynamics of the chemical bond, the interplay between theory and experiment, and bonding and dynamics, is illustrated in a number of articles. George Porter gives an overview of flash photolysis, the breakthrough that opened a window on chemical intermediates; here, Porter describes chemistry in microtime as well as the progress made over the last 40 years. In his unique style, John Polanyi discusses the transition state—a concept with roots in the

Polanyi family!—in chemical reactions and the insight and the surprising lessons one learns in studying elementary chemical reactions. Dudley Herschbach addresses a central problem for chemical bonding: the interplay between structure and dynamics. With enthusiasm, Herschbach overviews the past, highlights Pauling's influence, and projects into the future. In the last chapter, Bernstein (with the editor) introduces femtochemistry, chemistry on the femtosecond (10^{-15} second) time scale, and describes the ultrafast laser techniques developed for viewing, in real-time, the dynamics of the chemical bond.

This volume is an outgrowth of a symposium during Caltech's centennial year, held on a very special occasion—Linus Pauling's 90th birthday on February 28, 1991. The lectures were presented to a large audience (about 1,000 people) and the idea was for every contributor to provide an overview of a field, describing past achievements and future directions. I would like to take this opportunity to thank all the speakers, who were highly supportive and enthusiastic. The talks were wonderfully presented, were of the highest scientific quality, and kindled great interest among students, faculty, and laymen in the history of these scientific ideas and developments. Sadly, Dick Bernstein, who was scheduled to take part in the event, died on July 8, 1990. The last chapter of this book is on the subject that Dick intended to cover.

I am pleased to have been involved in the organization of this symposium and with a previous symposium celebrating Linus' 85th birthday. Linus and Caltech are one institution, only 10 years apart in age, and the bond between them is very strong. Since our first celebration (see photo, next page) in 1986, we now have at Caltech the Linus Pauling Lectureship, the Linus Pauling Professorship, and the Linus Pauling Lecture Hall. There was a distant period after Linus left Caltech in 1964, but, as Linus has told me on several occasions, he has the highest admiration for the California Institute of Technology, and the respect is mutual. In recent years, our Division of Chemistry and Chemical Engineering and the Institute have been very happy to see one of the most distinguished members of our faculty with us again!

I wish to thank the Presidents (Thomas Everhart and Marvin Goldberger) and the Provosts (Paul Jennings and Rochus Vogt) of Caltech for their support of the idea and for their contributions to the two events. The enthusiasm and support from Sunney Chan, Chairman of

Linus Pauling, crowned the "Pharaoh of Chemistry" in 1986 by Ahmed Zewail.

the Caltech Centennial Committee, is greatly appreciated. In planning the 1986 and the 1991 symposia, I had a real partner in Fred Anson. Fred and I spent many lunches together and devoted hours to the discussion of many details; his enthusiasm, sincerity, and fairness were qualities essential to the success of these historical events.

The contribution of Jane Ellis and everyone else at Academic Press was substantial—her unbelievable energy was transmitted to me either by phone or fax; Jane is certainly the "queen of faxes!" Every fax was detailed, thorough, and, above all, contained dozens of questions and requests. The quality of the book reflects the quality of Jane's editorship. In Chapter One, I called on the expertise of Dr. Stan Samson and, with his help, the figures were constructed. The photos in Chapters Four and Five are courtesy of the Caltech Archives and some are from our own collection of Linus Pauling visits. One of the photos (Professor Sommerfeld, in Chapter Five) was kindly provided by Linda Kamb, Linus' daughter.

Ahmed Zewail
Pasadena, California
October 1991

THE CHEMICAL BOND

STRUCTURE AND DYNAMICS

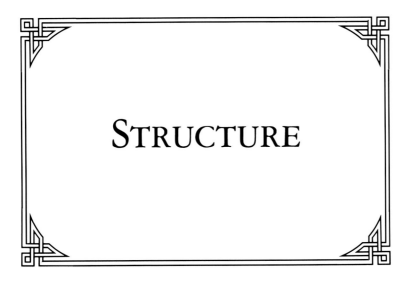

STRUCTURE

1

X-Ray Crystallography and the Nature of the Chemical Bond

Linus Pauling
Linus Pauling Institute of Science and Medicine

One evening in February 1912, P. P. Ewald, a student in Arnold Sommerfeld's Institute of Theoretical Physics in the University of Munich, came to Max von Laue to ask for advice about how to attack a problem given to him by Sommerfeld, that of interaction of electromagnetic waves with a space lattice of scatterers. While Max von Laue was thinking about this problem, he had the idea that x-rays might be electromagnetic waves with wavelength of the order of magnitude of the distance between layers of atoms in crystals. He got two young experimental physicists, Friedrich and Knipping, to carry out an experiment in which a beam of x-rays passed through a crystal of copper sulfate pentahydrate, and observed spots, diffraction maxima, on the photographic plates set up around the crystal. Laue then developed the Laue equations, relating the direction in which diffraction maxima occur to the dimensions of the unit of structure of the crystal. A few months later W. Lawrence Bragg, a student in Cambridge University, while preparing a talk on the Laue discovery, developed a simple equation, the Bragg equation, to describe the diffraction maxima. His father, William H. Bragg, set up an apparatus to test the Bragg equation, and in 1913 Lawrence Bragg and his father published several papers reporting not only the wavelengths of x-ray lines but also the structures of a number of crystals, diamond, sodium chloride, fluorite, and others. This was the start of x-ray diffraction and the determination of the structure of crystals.

FIGURE 1. The structure of NaCl as developed by Barlow, 1898, fourteen years before the discovery of x-ray diffraction.

In 1913 I was 12 years old. I was spending part of my spare time reading books on mineralogy and trying to understand the properties of minerals in relation to their chemical composition. In 1914 my interest became broader, to include the whole field of chemistry. Later, around 1920, I became interested in metals and intermetallic compounds, and I attempted, without success, to make single crystals of iron by electrolytic deposition in a magnetic field.

I consider my entry into the field of x-ray crystallography, nine years after it had been developed, to be just about the most fortunate accident that I have experienced in my life. I had become interested in the question of the nature of the chemical bond, after having read the 1916 paper on the shared-electron-pair chemical bond by G. N. Lewis and the several 1919 and 1920 papers by Irving Langmuir on this subject. Then, in the spring of 1922, when I was appointed a teaching fellow in the California Institute of Technology and had accepted the job, Arthur Amos Noyes wrote to me to suggest that when I came to Pasadena in the fall I should work with Roscoe Gilkey Dickinson, who was then a National Research Council fellow, on the determination of the structure of crystals by x-ray diffraction. He suggested that I get a copy of the book *X-Rays and Crystal Structure* by W. H. and W. L. Bragg and read it, which I did during the summer, and when I arrived in Pasadena toward the end of September 1922, Roscoe Dickinson began teaching me the techniques of x-ray diffraction.

X-ray diffraction had been begun in Pasadena in 1917 by Lalor Burdick, who had spent some time with W. H. Bragg after having got his Ph.D. in chemistry in Switzerland. Dickinson came with Noyes from M.I.T. in 1918, continuing the work, and receiving his Ph.D. in 1920 as the first Caltech Ph.D.

There were several new graduate students in chemistry at C.I.T. in 1922, and I can only surmise why Noyes decided that I should work with Dickinson on x-ray diffraction. The assignment may have been the result of a statement of my interest that I made in my letter of application for the teaching fellowship. It was a fortunate decision for me, in that even in those early days the technique was very powerful and, through the determination of interatomic distances, bond angles, ligancies, and other structural features it had important consequences for the question of the nature of the chemical bond.

In those early days it was difficult to determine the structure of a crystal with more than two or three parameters locating the atoms, in addition to the dimensions of the unit cell. It was soon recognized that in a crystal with high symmetry the number of parameters may be small, because atoms often were located on symmetry elements or at their intersections. Cubic crystals, because of their large number of symmetry operations, were the ones that were usually most easily susceptible to attack. In fact, Dickinson and a senior student, Albert Raymond, had just completed the determination of the first organic substance to have its structure determined. This was hexamethylenetetramine, $C_6H_{12}N_4$, which forms cubic crystals, such that the carbon atom is fixed by one parameter and the nitrogen atom by one parameter.

At Dickinson's suggestion, I began searching the literature, especially the several volumes of Groth's *Chemische Kristallographie,* to find cubic crystals that might be worth study. During the first two months I prepared several inorganic compounds and made crystals of them, in total number 14, and made Laue photographs and rotation photographs of each of them. They all turned out to be so complicated as to make it unwise to try to determine their structures. Dickinson then gave me a specimen of molybdenite, a hexagonal crystal, and helped me to solve the problems that arose in the determination of its structure. It turned out to be quite interesting; not only was the value of the molybdenum–sulfur distance of interest to me, but also the fact that the molybdenum atom, surrounded by six sulfur atoms, had these atoms at the corners of a triangular prism rather than a regular octahedron.

As I recall, I was interested in both the interatomic distance and the unusual coordination polyhedron, but I had no hope of developing an understanding of these structural features. I had begun collecting experimental values of interatomic distances in crystals, stimulated by a paper by W. L. Bragg, published in 1922. Bragg had formulated a set of atomic radii, starting with 1.04 Å for sulfur, half the distance between the two sulfur atoms in pyrite, with other values obtained by the assumption of additivity from observed interatomic distances. The trouble with the Bragg radius was clear from the molybdenite structure, where there are close-packed layers of sulfur atoms in juxtaposition to one another, with the sulfur–sulfur distance 3.49 Å, rather than

2.08 Å. The idea that an atom of an element could have several radii, a van der Waals radius, an ionic radius, a single-bond covalent radius, a double-bond covalent radius, a metallic radius, and so on, had not yet been developed. During the next few years, however, Wasastjerna developed a set of ionic radii and V. M. Goldschmidt developed a set of radii that was a sort of combination of covalent radii and metallic radii. It was clear that much additional information about the nature of the chemical bond was being provided and was going to be provided in the future by the determination of the structure of crystals by the x-ray diffraction method.

I shall not say much about the next decade, during which many interesting crystal structures were determined and some general principles about these structures were developed, especially in relation to the silicates, which in the course of a decade changed from being about the most poorly understood minerals to the best understood minerals.

In 1934 the transition from early x-ray crystallography to modern x-ray crystallography was begun by the discovery of the Patterson diagram by A. L. Patterson. Use of the Patterson diagram permitted a straightforward attack on the determination of the structure to be made for many crystals. I remember that in 1937 Robert B. Corey and I decided together that structures should be determined of crystals of amino acids and simple peptides, as a method of attack on the protein problem. At that time no correct structure for any of these substances had been reported. Structure determinations had been made for some related substances, simple amides, and the theory of the chemical bond had been developed to such an extent as to permit the conclusion to be reached with confidence that the peptide group (the amide group) in polypeptides should be very closely planar, and there was also reliable knowledge about the expected interatomic distances, including the N—H—O bond length. I had tried to use these structural elements in predicting ways in which polypeptide chains might fold, with the formation of hydrogen bonds, and in 1937 had decided that my efforts had failed, and that probably there was some structural feature yet to be discovered about proteins.

Eleven years later, when the alpha helix was discovered, about a dozen of these structures had been determined, all of them in the Gates and Crellin Laboratory of Chemistry. Verner Schomaker has pointed out to me that every one of these structure determinations made valuable use of Patterson diagrams. In 1948 I recognized, however, that my 1937 idea that there might be something new and surprising about amino acids and peptides was wrong. The structural features were the same in 1948 as in 1937.

During the period of 20 years, beginning in 1934, a number of investigators, including Patterson himself, made further contributions to the solution of the problem of direct determination of the structures of crystals from the x-ray diffraction pattern. Among these contributors were D. Harker, E. W. Hughes, V. Schomaker, W. H. Zachariasen, H. Hauptman, and J. Karle. The development of modern computers played an important part in this process. The result is that now x-ray techniques can be used to locate tens of thousands of atoms in a unit cell of a crystal. The time required to determine an only moderately complicated structure has been decreased from years to perhaps an hour. A tremendous amount of information bearing on the question of the nature of the chemical bond has been obtained. I have followed this development with much interest, but also with considerable disappointment. The disappointment is the result of the fact that, so far as I have been able to see, the more recent investigations have not led to the discovery of new structural principles. Some refinements might be made in the principles expressed in the third edition of my book *The Nature of the Chemical Bond,* 1960, which represent only rather minor changes from those in the first edition, 1939, but these refinements are minor.

There is one branch of structural chemistry that needs further development. We still do not have a good theory of structural chemistry of metals and intermetallic compounds, even though structure determinations have been made for 24,000 intermetallic compounds, representing about 4,000 distinct structures. There is still the possibility that some surprising discoveries will be made in this field.

I miss the old days, when nearly every problem in x-ray crystallography was a puzzle that could be solved only by much thinking.

Computers do not think, and the many errors in published structure determinations, such as the more than 100 pointed out by Dick Marsh, suggest that the younger x-ray crystallographers also do not think.

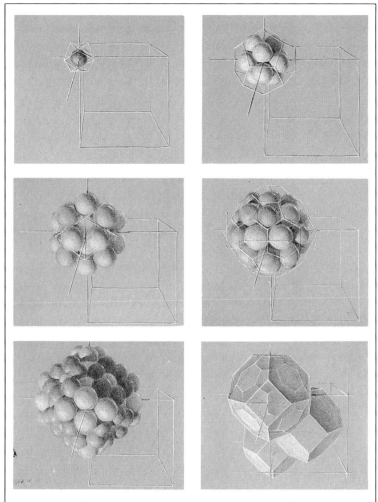

FIGURE 2. The crystal structure of the intermetallic compound $Mg_{32}(Al, Zn)_{49}$. The successive shells of atoms about the point $(0, 0, 0)$ and the sharing of the outermost atoms between neighboring groups. Adapted from G. Bergman, J. Waugh, and L. Pauling (1957), *Acta. Cryst.* **10,** 254. An example of work published from the California Institute of Technology.

An example of the usefulness of not having a computer is provided by my work during the last six years on the problem of the nature of the so-called icosahedral quasicrystals, which were discovered about six years ago by Shechtman and his collaborators, when they rapidly quenched molten alloys of aluminum with manganese and other metals. Electron-diffraction photographs of these alloys showed that they possessed fivefold, threefold, and twofold axes in the relative positions corresponding to an icosahedral point group. These investigators and many later ones assumed that the point-group symmetry extended down to close to the atomic scale, and that the alloys consisted of individuals with a sort of icosahedral space-group symmetry, which had not been accepted by crystallographers. Several remarkable theories about quasicrystals were developed, such as to permit them to be described as constituting a new form of matter. All of these theories involved some sort of randomness of structure, in order to take care of the problem that a fivefold axis of symmetry is incompatible with translational identity operations.

A few physicists and crystallographers, including me, pointed out, however, that there is the possibility that the appearance of icosahedral point-group symmetry might be the result of icosahedral twinning of cubic crystals or crystals with other conventional point-group symmetry. During the past two or three years, however, very few papers, other than my own, along this line have been published. Commentators have pointed out that my attitude with respect to the quasicrystals is rather different from my attitude with respect to other problems, in that usually I have been considered very radical, whereas in this matter I represent conservative crystallography.

This problem has turned out to be a very difficult one. I myself have made several starts at explanations, which usually were in the right direction but needed to be revised as new information and new ideas came into existence. I made all of the extensive calculations involved in my work without use of a computer, but with use of a small hand calculator. As I think back on the preceding years, I realize that with each small calculation I thought a bit about the nature of the calculation, and sometimes during this process had an idea about how to attack the problem in a somewhat different way. I am sure that if I had been relying on a computer to make most of the calculations, some of these ideas, which have in fact turned out to be important, would not have occurred to me.

I think that the evidence now that icosahedral quasicrystals are icosahedral twins of cubic crystals composed of very large icosahedral complexes of atoms is so strong that it should be convincing to everybody in this field. I am sure that the icosahedral quasicrystals that I have studied are in fact icosahedral twins of cubic crystals. Whether or not any alloys will ever be found that show icosahedral space-group symmetry as well as icosahedral point-group symmetry remains open.

Six years ago I presented my first argument to the American Crystallographic Association in its meeting in Stanford University. Some excellent high-resolution electron micrographs had been published, showing what appeared to be lines of atoms extending in several directions and indicating fivefold axes or pseudo-axes of symmetry. The resolution of the photographs is such that if fivefold axes really were present, their presence would show that the symmetry extends down to atomic dimensions and could be described as point-group symmetry (or approximate symmetry). The pentagonal angle is 108°. If, on the other hand, the photographs represented layers of atoms in the 1.1.0 plane of a cubic crystal, the angle would be the tetrahedral angle, 109.47°. I had carefully measured this angle on one of the excellent electron diffraction photographs. I reported that my measurement gave 109.5°, probably reliably enough to rule out 108°, and accordingly showed that the quasicrystal consisted of cubic crystallites, icosahedrally twinned. There was, however, some skepticism about the reliability of my measurement, in that I might have been biased by my conviction that the quasicrystals consisted of twinned cubic crystallites. I then asked my associates Zelek S. Herman and Peter J. Pauling to measure, independently of one another, several of the published high-resolution electron-micrograph patterns, and I again measured several of them. We obtained similar results, the average of our measurements of the angle being 109.61°, with probable error of the mean 0.11°. This value supports the hypothesis that the quasicrystal consists of cubic crystallites. The micrographs that were measured extended over an area usually about 200 Å in diameter. These results were published in the *Comptes rendus of the Academy of Sciences, Paris,* in 1988.

One of the arguments that was proposed to rule out the cubic-crystallite hypothesis is that the relative spacing of some of the diffraction maxima corresponds closely to the golden number $1.6180\ldots$, rather than to ratios of integers (the Fibonacci numbers), these ratios, starting with Fibonacci number 3 and extending to 5, 8, 13, 21, being $1.667, 1.600, 1.625, 1.615, \ldots$ One opportunity to test these ratios was provided by the careful diffractometer measurements made by Budai and his collaborators in the General Electric laboratories. They had prepared an oriented quasicrystal with composition approximately Al_5Mn by starting with a single crystal of aluminum and bombarding it with manganese ions. This process of ion implantation led to the oriented quasicrystal, from one face of which a careful diffractometer measurement could be made. I measured the positions of the peaks on their published curve, and found the first three ratios to lie between the golden number and the Fibonacci ratios. Dr. Budai kindly provided me with the accurate values of the Bragg angle measured with the diffractometer, so that I could repeat the calculation without fear of bias on my part. It was found that the first three ratios, corresponding to Fibonacci numbers 3, 5, 8, and 13, were in fact intermediate between the Fibonacci ratios and the golden number, so that they moved in alternation from values lower and higher than the golden number. They were in fact only about 15% of the way from the golden number to the Fibonacci ratio, but the displacement was definite enough to indicate that the diffraction maxima correspond to an overlap of the peaks on two form factors: the form factor for the icosahedral complexes, which would have its maxima in ratios equal to the golden number, and the form factor of a cubic lattice, which would have its maxima at the Fibonacci ratios.

Additional evidence for the hypothesis of twinned cubic crystallites was provided by electron micrographs of clusters of crystallites prepared by Agnes Csanády and her collaborators. Dr. Csanády had isolated small clusters of crystallites of the Al_4Mn quasicrystal by quenching alloys containing an excess of aluminum and then removing the excess solvent by electrosolution, leaving the quasicrystals. With a considerable excess of aluminum, individual clusters were obtained, consisting of 20 crystallites presumably surrounding a central one, which served as seed for the icosahedral twinning. The individual crystallites could be seen to be cubes, with somewhat rounded edges

FIGURE 3. The quasicrystals of Al₄Mn isolated by Csanády *et al.*
(1990), *Symmetry* **1**, 75, reprinted by permission of VCH Publishers, Inc., 220 East 23ʳᵈ St., New York, N.Y., 10010.

and corners. They are differently oriented, corresponding to the different ways of having a threefold axis and planes of symmetry in common
with an icosahedron. It is my opinion that Dr. Csanády's electron micrographs provide strong evidence that icosahedral quasicrystals (at
any rate the Al₄Mn quasicrystals) are in fact icosahedral twins of cubic
crystals.

It can be seen from the electron micrograph of this cluster of cubic crystallites that if the cluster had grown to such an extent that the crystallites came into juxtaposition with one another, their cubic faces on the outside would coalesce into the 20 faces of the pentagonal dodecahedron. In fact, in preparations made with a small amount of excess aluminum as solvent, the quasicrystal separation was seen to consist of apparent pentagonal dodecahedra. For other crystals, too, preparations that seem to be single crystals have a face development corresponding to polyhedra with icosahedral symmetry. All of these observations are compatible with the idea that the quasicrystals are twins of cubic crystals.

I have made over the past six years several attempts to analyze electron-diffraction photographs, x-ray diffraction patterns, and neutron diffraction patterns of quasicrystals, but with only partial success. About a year ago, while carrying out one of these attempts at analysis, I realized that there is something special about the twofold-axis electron-diffraction photographs of an icosahedrally twinned aggregate of cubic crystallites. The fact that it took me several years to reach this conclusion shows that sometimes the solution to a problem may come only after a long sequence of efforts to solve it. One problem is that an electron-diffraction photograph of an oriented quasicrystal taken with the electron beam along the fivefold axis would, on the twinned-crystallite basis, be the superposition of 10 diffraction photographs, and a similar complexity would hold also for photographs taken along the threefold axis. With the beam along the icosahedral twofold axis, however, the only planes in a position to diffract are the planes in the form h.h.l of two of the 20 crystallites. These two cubic crystallites have this form in the plane of the incident electron beam, so that, because of the very small angle of diffraction, the planes in this form would be able to diffract. For the other 18 orientations, the relation between the direction of the incident beam and the crystallographic axes of the crystallites is irrational, and only by a rather rare accident would a plane be able to diffract the electron beam. Having recognized this fact, I looked carefully at the published twofold-axis electron diffraction photographs of many different alloys. I found that the diffraction maxima lay on layer lines, both horizontally and vertically, as well as on the diagonal layer lines at the angles characteristic of an oriented complex with icosahedral symmetry. I was astonished to see that the layer lines, which were in fact well defined, required that the edge of the unit cube be very large, far larger than I had considered in my earlier efforts to analyze powder patterns and other diffraction patterns. Moreover, the twofold-axis electron-diffraction photographs of different alloys were found to differ from one another in ways recognizable on inspection, with three different patterns easily recognizable. All of these patterns showed the characteristic expected for large atomic complexes with icosahedral symmetry. The layer lines for one of the patterns were found to correspond to a face-centered cubic structure, with edge about 66.3 Å, and with about 29,400 atoms in the cube. This structure was found for the quasicrystal Al_5Mn. Another quasicrystal, $Al_{13}Cu_4Fe_3$, was indexed as having a body-centered cubic unit with edge 52.0 Å containing about 9,960 atoms. The third structure, found for the quasicrystals Al_6CuLi_3, was indexed as having a simple cubic unit with edge 57.8 Å, containing about 11,700 atoms.

I have made some tentative suggestions about the atomic arrangements in these three kinds of quasicrystals. For the first two I assume that there are complexes of nearly 5,000 atoms, with icosahedral symmetry, arranged in cubic closest packing in the first structure and in the body-centered arrangement in the second structure. For the third structure I assume that there are eight icosahedral complexes, each of about 1,500 atoms, arranged with their centers in the beta-tungsten simple cubic positions.

I have now made one more effort to gather additional evidence to support the twinned-cubic-crystallite structure of the icosahedral quasicrystal. Two years ago T. Egami and S. J. Poon sent me pulsed neutron powder diffraction data for four quasicrystals. The diffraction patterns show a score or more of rather strong diffraction maxima and a very great number of quite weak peaks. I agreed with the investigators that the very weak peaks were probably just the result of instrumental fluctuations, and my first efforts at analyzing the powder patterns consisted of considering only the interplanar distances of the stronger peaks. A few months ago, however, I realized that the large values of the edge of the unit cubes that had turned up in the study of the twofold-axis electron-diffraction photographs would permit the indexing also of the numerous weak peaks. Moreover, three of the alloys for which the pulsed-neutron diffraction data had been sent to me had composition differing little from Al_6CuLi_3, for which I had proposed a simple-cubic structure with cube edge 57.8 Å.

With such a large unit of structure, and with a little uncertainty about the value of the edge of the unit cube, the only peaks that can be reliably indexed on a powder pattern are those at rather small Bragg angles. It was found that for all three of these alloys the indexing of the low-angle weak peaks could be successfully carried out, the values of the edge of the unit cube turning out to be 58.4, 58.5, and 58.6 Å—a little larger, as expected from their compositions, than the value found for Al_6CuLi_3, 57.8 Å. The fourth alloy for which the pulsed-neutron data were available, Pd_3SiU, was also found to have a simple-cubic structure, with cube edge somewhat smaller, 56.2 Å, corresponding to the somewhat smaller atomic radii of the elements. All three alloys have about 11,600 atoms in the unit, probably representing eight icosahedral complexes of 1,500 atoms each.

I am now satisfied that the solution to the puzzle of the existence of the icosahedral quasicrystals has been found, and that I may from now on devote my time to other pursuits.

2

THE SIGNIFICANCE OF THE HYDROGEN BOND IN PHYSIOLOGY

Max F. Perutz,
MRC Laboratory of Molecular Biology

I would like to describe the influence that Linus Pauling had on me as a scientist, even though I met him only sporadically and never actually worked in his laboratory.

I read chemistry in Vienna, where the departmental chairman lectured on the whole of inorganic and organic chemistry. He taught us the reactions of the elements and the organic syntheses. I wrote down what he said, or rather my girl friend who knew shorthand did, and I learnt it by heart.

Later I became a graduate student in x-ray crystallography at Cambridge, but nobody there told me that chemistry could be understood, rather than merely memorized. At Christmas 1939 another girl friend, an English one, gave me a book token. I used it to buy Linus Pauling's *The Nature of the Chemical Bond*.

In that first, 1939 edition of *The Nature of the Chemical Bond,*
Pauling introduces the hydrogen bond in the following words:

> Although the hydrogen bond is not a strong bond (its bond energy,
> that is the energy of the reaction $XH{=}Y{\rightarrow}XHY$, being only about
> 5 kcal/mole), it has great significance in determining the proper-
> ties of substances. Because of its small bond energy and the small
> activation energy involved in its formation and rupture, the hydro-
> gen bond is especially suited to play a part in reactions occurring
> at normal temperatures. It has been recognized that hydrogen
> bonds restrain protein molecules to their native configurations,
> and I believe that as the methods of structural chemistry are
> further applied to physiological problems it will be found that the
> significance of the hydrogen bond for physiology is greater than
> that of any other single structural feature.

Pauling's estimate of the energy of hydrogen bonds is exactly the
same as that calculated 35 years later from Hagler and Lifson's (1974)
van der Waals and electrostatic contributions to the lattice energies of
amide crystals.

In 1948, Pauling suggested what he thought might be the best molecule to test his prediction experimentally. He wrote: "Hemoglobin is
one of the most interesting and important of all substances, and even
the great amount of work that would be needed for a complete determination of its structure, involving the exact location of each of the
thousands of atoms in its molecule, would be justified" (Pauling, 1949).
I took this admonition to heart, but it took me another 30 years to do
the job (Fig. 1).

What stabilizes the unique fold of the polypeptide chains in hemoglobin? Physical chemists who studied this problem before the elucidation of protein structures by x-ray analysis were struck by the weak-
ness of the association between both neutral and ionized hydrogen
bond-forming groups in water. They argued that hydrogen bonds and
salt bridges could not provide net stability to native proteins in water,
because in the unfolded molecule the same hydrogen bond-forming
groups could make bonds of similar energy with the solvent. Kauzmann (1959) therefore suggested that the stability of globular proteins
should come mainly from nonpolar bonding, that is, from the entropy

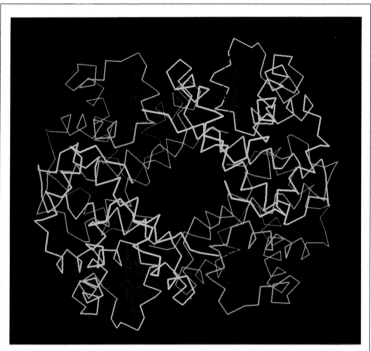

FIGURE 1. α-Carbon skeleton of human deoxyhemoglobin.

gained by the surrounding water on burying the hydrophobic side chains in the interior of the protein. Experiments on the partition coefficients of hydrocarbons between nonpolar solvents and water have shown that this entropy gain is equivalent to a free energy loss $\Delta G^\circ = -24$ cal/mol/Å² of buried surface. This figure can be used to draw up a balance sheet of the opposing forces in globular protein. It shows that the free energy of hydrophobic bonding is not nearly sufficient for stability. Where does the remaining energy come from? Firstly, the interiors of proteins are packed nearly as densely as crystals, so that dispersion forces make a large contribution. Secondly, hydrogen bonds do in fact make large, and in some proteins probably dominant, contributions to stability; this has been demonstrated experimentally in hemoglobin, which consists of four subunits joined by secondary valency forces. Included among these are two internal hydrogen bonds between the phenolic hydroxyls of tyrosines and the carboxylates of aspartic acids. In one mutant hemoglobin, one of the tyrosines is replaced by a phenylalanine (Asakura *et al.*, 1976); in another, one of the aspartates is replaced by an asparagine. The Tyr→Phe replacement destabilizes the interaction between the subunits by the equivalent of 3 kcal/mol, the Asp→Asn replacement by 1.7 kcal/mol.

In 1964, when Pauling had seen the position of the distal histidine at hydrogen-bonding distance from the heme-linked water molecule in Kendrew's structure of ferric or metmyoglobin, he predicted that it would form a hydrogen bond with the bound oxygen in both myoglobin and hemoglobin (Pauling, 1964). His prediction was based on valence bond theory: One oxygen atom would form a double bond with the iron atom, giving it a positive charge, while the second oxygen atom would carry a negative charge, thus attracting a hydrogen bond from the $N_\varepsilon H$ of the histidine.

Seventeen years later, my young colleague Simon Phillips solved the structure of oxymyoglobin at 1.6 Å resolution and found the $N_\varepsilon H$—O distance to be 2.8 Å (Phillips, 1980); this distance is consistent with a hydrogen bond, but it does not prove it. The distal histidine has a pK_a of 5.5. At physiological pH, therefore, its imidazole nitrogens carry only one hydrogen, which could be on either N_ε or N_δ. If the hydrogen were on N_δ, facing the external solvent, then the naked N_ε would be in van der Waals contact with the oxygen atom, with an N–O distance of 3.4 Å. The difference of 0.6 Å was larger than the experimental error, but even so we decided to try and verify the hydrogen bond by neutron diffraction. Phillips took an enormous crystal of oxymyoglobin to Brookhaven, exchanged its water of crystallization against D_2O, and with Benno Schoenborn's help, collected neutron diffraction intensities for 14,000 reflexions. The results showed a deuteron sandwiched between the iron and the oxygen, providing conclusive evidence for the hydrogen bond Pauling had predicted (Fig. 2) (Phillips and Schoenborn, 1981).

How much does this bond contribute to the binding energy of an oxygen molecule to myoglobin? Gene technology made it possible to answer this question. Hemoglobin and myoglobin can now be produced in any quantity in coli bacteria or in yeast, and amino acids can be substituted as desired, simply by replacing the appropriate codon in the c-DNA.

The replacement of the distal histidine by glycine leaves the oxygen affinity and kinetic constants of the β-subunits unchanged within error; it diminishes the oxygen affinities of myoglobin 14-fold, and that of the α-subunits eightfold. In myoglobin, the reduction is equivalent to stabilization of the bound oxygen by hydrogen bonds with the histidines by the equivalent of 1.4 kcal M^{-1}, about half the energy of the hydrogen bond linking two amides (Rohlfs *et al.*, 1990); the other half may be accounted for by the loss of rotational entropy of the histidine side chain making the bond. X-ray analysis shows that the distal histidine blocks access to the heme pocket. Neither O_2 nor CO can enter or leave unless the side chain of the distal histidine swings out of the way, which it can do only by elbowing the helix E, to which it is attached, away from the heme. Thus, oxygen transport relies on the dynamics of the globin.

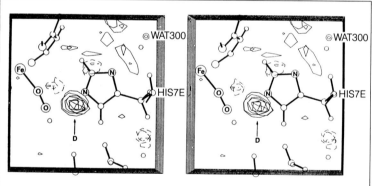

FIGURE 2. $|F_o|-|F_c|$ neutron difference density map in a slab 4 Å thick centered on the plane of the E7 imidazole ring. The refined model is superimposed, showing His E7, FeO_2, and part of the heme. Contours are ± 0.35, 0.55, 0.75 Fermi per Å, with negative ones shown as broken lines. A strong positive peak indicates the presence of deuterium bonded to N'. (Reproduced by permission from S. E. V. Phillips and B. Schoenborn, *Nature* **292**, 81–82, Copyright © 1981 Macmillan Magazines Ltd.)

Supposing there were no barrier and the oxygen could enter freely through a hole of 2.6 Å radius, Szabo (1978) calculated that this would bring the on-rate for oxygen to 50×10^7 s^{-1}M^{-1}, 36 times faster than the observed rate; if the barrier to the heme pocket is removed by replacing the distal histidine by glycine, the on-rate rises 10-fold and is now only 3.6 times slower than the calculated diffusion rate.

At first sight, the heme pockets of myoglobin and hemoglobin look alike, but this picture is deceptive. In the α-heme pocket, the distance from N_ε to oxygen is 2.6 Å, the same within error as in myoglobin and consistent with the formation of a hydrogen bond, but in the β-heme pocket the distance is 3.5 Å, indicative of no more than a van der Waals contact. Consistent with the x-ray observations, replacement of the distal histidine by glycine reduces the oxygen affinity by the equivalent of 1 kcal/mol in the α-subunits, but it leaves the oxygen affinity of the β-hemes unchanged (Mathews et al., 1989). Their on-rate for oxygen is so near the diffusion limit that the replacement produces no further acceleration. From this observation we must infer that in native hemoglobin the histidine swings in and out of the pocket at least 200 million times a second.

Computer simulations of the molecular dynamics of the exit of carbon monoxide from the interior of myoglobin, described by Elber and Karplus (1990), suggest that there may be several alternative pathways, in addition to that via the distal histidine, even though the latter is the most direct. Experimental evidence in its support comes from a recent crystal structure determination of ethylisocyanide myoglobin that shows the side chain of the distal histidine in two alternative positions, either in or out of the heme pocket, exactly as it would have to move to admit or release ligands (Johnson et al., 1989) (Fig. 3). The dynamic movements of the heme pocket are attested by NMR studies (Dalvit and Wright, 1987) showing that phenylalanines CD1 and CD4, which wedge the heme into its pocket and are packed tightly between the heme and the distal helix E, flip over at rates faster than 10^4 s^{-1}. They can do so only if the entire heme pocket breathes fast.

Only ferrous heme combines reversibly with molecular oxygen. Free heme is quickly oxidized by contact with air, but the globin pocket protects it from oxidation. How? In 1962 Pauling suggested that the distal histidine "assists in preventing oxidation to ferrihemoglobin"

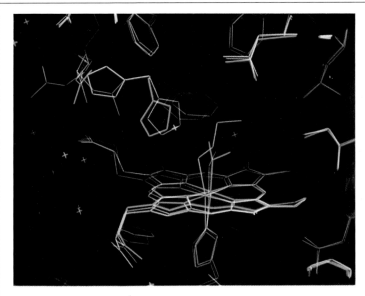

FIGURE 3. Superposition of the structures of oxymyoglobin (red), carbonmonoxymyoglobin (blue), and myoglobin–ethyl isocyanide (green), showing a view of the heme pocket. His E7 can be seen (top center) in several conformations: swung down near the bound dioxygen and carbon monoxide and swung out of the heme pocket and away from the bound ethyl isocyanide. The most occupied position of the His E7 side chain in the myoglobin–ethyl isocyanide structure is rotated by 102° around the C_α—C_β bond relative to its position in the deoxy structure. (Reproduced by permission from K. A. Johnson, J. S. Olson, and G. N. Phillips, (1989), *J. Mol. Biol.* **207**, 459–463.)

(Zuckerkandl and Pauling, 1962). Recently, that prediction was also tested by gene technology.

Springer *et al.* (1989) have studied the protection of the heme iron from oxidation by replacing the distal histidine in sperm whale myoglobin by 10 different amino acid residues and shaking the deoxygenated myoglobin solutions in air in 75 mM K_2HPO_4 + 25 mM EDTA pH 7.0 at 37°C. All replacements reduced the oxygen affinity and accelerated autooxidation. Phe, Met, and Arg produced the smallest accelerations (~50-fold), and Asp the largest (350-fold).

How can these results be interpreted? Paradoxically, combination with oxygen protects the heme iron from oxidation, as can be shown by performing the same experiments at several atmospheres of pure oxygen. Apparently oxidation occurs in that fraction of molecules which are deoxygenated at any one moment. The larger that fraction is at the atmospheric oxygen pressure, the faster myoglobin autooxidizes. For example, replacement of the distal His by Phe reduced the oxygen affinity 170-fold, so that a larger fraction of myoglobin molecules will have remained deoxygenated at atmospheric oxygen pressure and therefore have become autooxidized. However, this can be only part of the explanation, because the replacement of His by Gly reduces the oxygen affinity merely 11-fold, yet accelerates autooxidation over 100-fold.

Autooxidation is catalysed by protons, hence the 350-fold acceleration by Asp. Protons are reduced at the heme iron and in their turn reduce oxygen in the solvent to superoxide ion. I suggest that the distal histidine protects the ferrous heme iron by acting as a proton trap. The distal histidine has a pK_a of about 5.5; at neutral pH it is protonated only at N_δ, which faces the solvent. Any proton entering the heme pocket of deoxymyoglobin would be bound by N_ε, and simultaneously N_δ would release its proton to the solvent. When the histidine side chain swings out of the heme pocket, the protons would interchange, restoring the previous state. No other amino acid side chain could function in this way. Evolution is a brilliant chemist (Perutz, 1989).

Hemoglobin is a two-way respiratory carrier: It carries oxygen from the lungs to the tissues, and it promotes the return transport of CO_2 from the tissues to the lungs indirectly. On releasing its oxygen, hemoglobin takes up the protons liberated by the reaction of carbon dioxide with water to form soluble bicarbonate; on taking up oxygen in the lungs, hemoglobin releases protons which turn bicarbonate into CO_2 that has a low solubility in water and can now be exhaled.

In 1932 Conant suggested that the uptake and release of protons might be accomplished by linkage of the heme iron to two histidines, one on either side of the hemes, that changed their pK_as on uptake and release of oxygen (Conant, 1932).

Conant's first prediction about the presence of the histidines was right, but his second prediction about their changing their pK_as was inconsistent with the absence of any proton uptake or release by myoglobin, which has the same heme-linked histidines as hemoglobin.

The true mechanism turned out to be different. The deoxy structure of hemoglobin is constrained by salt bridges which open on transition to oxyhemoglobin. One pair of these bridges is formed by histidines forming hydrogen bonds with aspartates. In oxyhemoglobin the histidine and the aspartate are far apart, and the histidine's imidazole has a pK_a of about 7. In deoxyhemoglobin the two side chains are brought into contact. The aspartate's negative charge raises the pK_a of the histidine to 8 and attracts a proton which gives the imidazole a net positive charge; the two side chains are now joined by a hydrogen bond (Perutz et al., 1987) (Fig. 4). The uptake of a proton by this histidine is

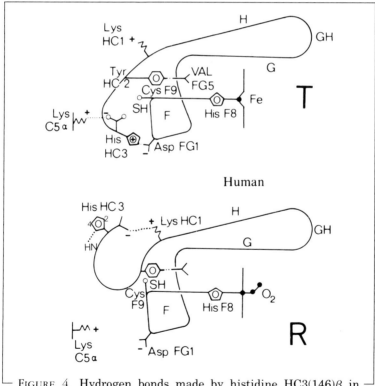

FIGURE 4. Hydrogen bonds made by histidine HC3(146)β in deoxy (T) and oxy (R) hemoglobin.

the most striking instance of the importance of the hydrogen bond in physiology, because without it, animals would not be able to get rid of their CO_2 fast enough to be able to run.

Some years ago, several colleagues asked me to help them in the design of drugs against sickle cell anemia. They would synthesize compounds and test their efficiency, while I would use x-ray crystallography to determine their binding sites in the hemoglobin molecule. In the course of this work, hemoglobin was found to bind a variety of drugs, and the study of their complexes with hemoglobin led to the formulation of general rules about the interactions between drugs and proteins. Our results suggested that the detailed stereochemistry of binding is likely to be governed by a tendency to maximize the sum of the energies of electrostatic interactions, for example, by aligning the drug relative to the protein so that the mutual polarizabilities of neighbouring groups are maximized (Perutz et al., 1986). After I had written down what I believed to be a new rule, I found that Pauling had already predicted it in another context 50 years earlier.

Pauling's 1939 edition of The Nature of the Chemical Bond says that "the free energy of interaction between a hapten and an antigen will be proportional to the sum of the polarizabilities of the atoms in contact."

I am glad to say that we did discover something which I believe he did not predict. In one of our complexes the amide of an asparagine pointed at the π-electrons of a benzene ring of the drug with what looked like a hydrogen bond, but our atomic coordinates were not accurate enough to allow us to decide whether the separation between them was shorter than the van der Waals distance (Fig. 5). Nevertheless my discussion of that possibility led a theoretician, Clare Woodward, to have a close look at one of the most accurately determined protein structures, that of the pancreatic trypsin inhibitor. This shows the benzene ring of a phenylalanine sandwiched between two amides, each of them closer than the van der Waals distance (Fig. 6) (Tüchsen and Woodward, 1987). Woodward's finding enabled Levitt to calculate the likely energy of such a bond, which he found to be about half of that of a hydrogen bond between two amides (Levitt and Perutz, 1988). A computer search by Burley and Petsko (1986) for similar bonds in proteins suggests that they make a significant contribution to the stability of their structures.

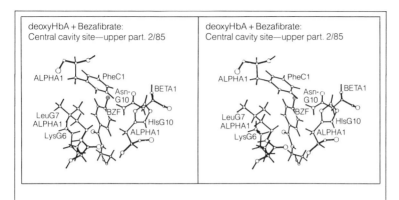

FIGURE 5. Stereoview showing the interaction of asparagine (Asn) G10 with the benzene ring of the drug bezafibrate bound to the central cavity of human deoxyhemoglobin. (Reproduced by permission from M. F. Perutz, G. Fermi, D. J. Abraham, C. Poyart, and E. Bursaux, *J. Am. Chem. Soc.* **108**, 1064–1078. Copyright © 1986 American Chemical Society.)

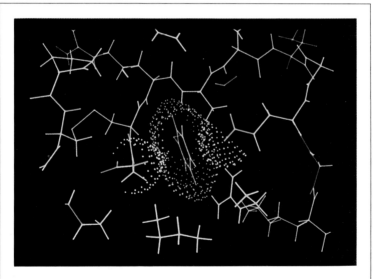

FIGURE 6. Interactions of amide groups with the benzene ring of a phenylalanine sandwiched between a main chain NH (left) and the NH_2 of an asparagine (right) in pancreatic trypsin inhibitor. (Courtesy Dr. Clare Woodward.)

Last year F. A. Cotton dropped in to see me. When I showed him our evidence, he pointed out that such hydrogen bonds had been found before and that an example of one originally published by McPhail and Sim (1965) was included in the second edition of Cotton and Wilkinson's *Advanced Inorganic Chemistry* (Fig. 7). This compound contains an intramolecular bond from an OH to a benzene ring, with a distance of 3.07 Å from the oxygen to the benzene plane. A computer search in the crystallographic data file by M. Levitt for NH–benzene interac-

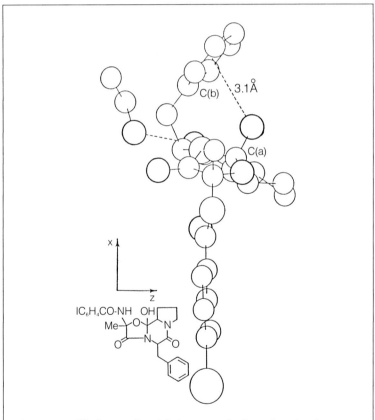

FIGURE 7. Hydrogen bond between a hydroxyl and a benzene ring in the crystal structure of a cyclic peptide. (Reproduced by permission from A. T. McPhail and G. A. Sim (1965), *Chem. Comm.*, 125–127.)

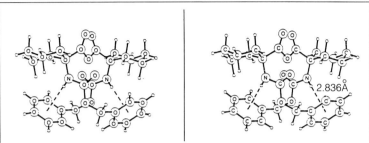

FIGURE 8. Stereoview of internal hydrogen bonds between two NH groups and two benzene rings in (benzyloxycarbonyl-1-aminocyclohexyl-1-carboxylic acid)$_2$-O. (Courtesy Dr. David Watson, University Chemical Laboratory, Cambridge, UK.)

tions led to the paper by Valle and others (1988) on the structure of a peptide that contains two intramolecular NH–benzene bonds with distances from N to the benzene plane of only 2.8 Å (Fig. 8). Both distances are much shorter than the sums of the van der Waals radii, which would be between 3.5 and 4.0 Å. These structures provide firm evidence that OH–C_6H_6 and NH—C_6H_6 bonds exist.

REFERENCES

Asakura, T., Adachi, J. S., Wiley, J. S., Fung, L. W. M., Ho, C., Kilmartin, J. V., and Perutz, M. F. (1976). *J. Mol. Biol.* **104**, 185–195.

Burley, S. K., and Petsko, G. A. (1986). *FEBS Letts.* **203**, 139–143.

Conant, J. B. (1932). *Harvey Lectures* **28**, 159.

Dalvit, C., and Wright, P. E. (1987). *J. Mol. Biol.* **194**, 313–327.

Elber, R., and Karplus, M. (1990). *J. Amer. Chem. Soc.* **112**, 9161–9175.

Hagler, A. T., and Lifson, S. (1974). *J. Amer. Chem. Soc.* **96**, 5327–5335.

Johnson, K. A., Olson, J. S., and Phillips, G. N. (1989). *J. Mol. Biol.* **207**, 459–463.

Kauzmann, W. (1959). *Advances Prot. Chem.* **14,** 1–62.

Levitt, M., and Perutz, M. F. (1988). *J. Mol. Biol.* **201,** 751–754.

Mathews, A. J., Rohlfs, R. J., Olson, J. S., Tame, J., Renaud, J.-P., and Nagai, K. (1989). *J. Biol. Chem.* **264,** 16573–16583.

McPhail, A. T., and Sim, G. A. (1965). *Chem. Comm.,* 125–127.

Pauling, L., in *Haemoglobin* (F. J. W. Roughton and J. C. Kendrew, eds.), pp. 57–66. Butterworths Scientific Publication, London, 1949.

Pauling, L. (1964). Nature **203,** 182–183.

Perutz, M. F. (1989). *Trends Biochem. Sci.* **14,** 42–44.

Perutz, M. F., Fermi, G., Abraham, D. J., Poyart, C., and Bursaux, E. (1986). *J. Amer. Chem. Soc.* **108,** 1064–1078.

Perutz, M. F., Fermi, G., Luisi, B., Shaanan, B., and Liddington, R. C. (1987). *Acc. Chem. Research* **20,** 309–321.

Phillips, S. E. V. (1980). *J. Mol. Biol.* **142,** 531–554.

Phillips, S. E. V., and Schoenborn, B. (1981). *Nature* **292,** 81–82.

Rohlfs, R. J., Mathews, A. J., Carver, T. E., Olson, J. S., Springer, B. A., Egeberg, K. D., and Sligar, S. G. (1990). *J. Biol. Chem.* **265,** 3168–3176.

Springer, B. A., Egeberg, K. D., Sligar, S. G., Rohlfs, R. J., Mathews, A. J., and Olson, J. S. (1989). *J. Biol. Chem.* **264,** 3057–3060.

Szabo, A. (1978). *Proc. Nat. Acad. Sci. USA* **75,** 2108–2112.

Tüchsen, E., and Woodward, C. (1987). *Biochemistry* **26,** 1918–1925.

Valle, G., Crisma, M., Toniolo, C., Sen, N., Sukumar, M., and Balaram, P. (1988). *J. Chem. Soc. Perkin Trans.* II, 393–398.

Zuckerkandl, E., and Pauling, L., in *Horizons in Biochemistry* (M. Kasha and B. Pullman, eds.). Academic Press, New York & London, 1962.

3

MOLECULAR RECOGNITION BETWEEN PROTEIN AND NUCLEIC ACIDS

Alexander Rich
Massachusetts Institute of Technology

The nucleic acids and the proteins are the two main polymeric materials of biological systems. Their molecular structures have evolved in association with each other, and it is apparent that they are completely interdependent in a functional sense. The nucleic acids contain genetic information. In their sequence of four nucleotides, they specify the information that is necessary for organizing and regulating the metabolic activities of a living cell. However, most of the machinery that carries out these activities is protein. The expression of the genetic information found in the nucleic acids is dependent upon the specificity of their interaction with proteins. Thus, the proteins regulate whether or not the genetic information in the nucleic acids will be expressed, and whether the processes of transcription or translation actually occur.

In recent years we have developed considerable understanding of the three-dimensional structure of proteins and nucleic acids, and we recognize that they occur with certain predominant structural motifs. These motifs are important in defining the nature of the interaction between the two macromolecules. Here we review information currently available on the nature of these interactions and discuss the manner in which our structural information limits and defines the nature of protein–nucleic acid interactions.

Linus Pauling has made monumental contributions in this field. These contributions span the entire gamut of protein structure and the way biological reactions are carried out. In 1946, he published an article (Pauling, 1946) dealing with the search for structure. There he discussed the fact that although we knew at that time a fair amount about the structure of small molecules, most structural features were unknown in the 10–1,000 Å range. Pauling focused on the fact that we needed to have knowledge about the modes of interactions between macromolecules where the outer part would be most important. He pointed out that this is likely to be the area that will govern most biological reactions. This statement made over 45 years ago is amply borne out by subsequent developments in this science. Now we know a fair amount about the structure of large molecules, proteins, and nucleic acids, and in studying their interactions, we are beginning to develop the kind of understanding that Pauling described over 45 years ago.

The Morphology of the Nucleic Acids

The most important structural motif found in the nucleic acids is that of the double helix, first described for DNA (Watson and Crick, 1953) and then extended to RNA as well. Structural studies of the nucleic acids have been carried out using a variety of techniques over the past four decades. In the beginning of this period, single-crystal analyses were carried out with individual nucleotides or nucleic acid bases. These began to provide us with information concerning the detailed geometry of the nucleic acid components. However, their macromolecular organization was studied in this earlier period predominantly through the method of fiber x-ray diffraction. The double helix is very well suited for that kind of analysis, since it is a long asymmetric structure that can be readily organized into an oriented fiber in which the helical axis is parallel to the fiber axis. The x-ray diffraction analysis revealed the gross features of the double helix. However, there is an important constraint inherent in fiber x-ray diffraction studies, since the resolution of the fiber patterns is limited, and this means that fine structural details cannot be resolved. In addition, fiber diffraction patterns are generally not "solved," that is, solved in the manner in which single-crystal x-ray diffraction patterns are solved through the elucidation of the phases of the diffraction pattern. Most of the double helical patterns were interpreted by means of a molecular model that was then modified to fit the structure as best as could be obtained. Since it was usually very difficult to obtain quantitative information from fiber diffraction patterns, the resultant solution had only limited reliability.

Nonetheless, out of this work there developed an appreciation of two different types of diffraction patterns exhibited by DNA, that is, an A pattern in which the DNA fiber was allowed to dry and the B pattern in which the DNA fiber was maintained in a moist condition. The A and B conformations differ from each other largely through the pucker of the sugar ring that is an integral part of the nucleic acid backbone. In the familiar B-DNA double helix, the ring has a C2' endo conformation in which the distance between the successive phosphate groups is approximately 6.7 Å. In the A conformation, the ring adopts the C3' endo pucker, which shortens the distance between successive phosphate groups by almost 1 Å. Because DNA is normally hydrated *in vivo*, the B conformation is considered as the normal or ground state of the molecule. However, the A conformation in DNA can also be observed. What is more important, the A conformation is the normal or lowest-energy conformation seen in the RNA double helix, since the addition of an oxygen in the 2' ribose position stabilizes the C3' endo conformation to a considerable extent.

Starting in the early 1970s, attempts were made to crystallize nucleic acid oligonucleotides so that they could be studied by single-crystal x-ray diffraction analysis. The first of these studies was carried out in 1973, involving the double helices formed by the ribonucleotides ApU (Rosenberg *et al.*, 1973; Seeman *et al.*, 1976) and GpC (Day *et al.*,

1973; Rosenberg *et al.*, 1976).[1] These fragments crystallized in an A conformation with Watson–Crick hydrogen bonds between the base pairs. Since the structures were carried out at the resolution of 0.8 Å, it was possible to define in great detail not only the conformation of the sugar phosphate backbone and the bases, but also the disposition of water molecules and ions surrounding the structure. This opened the era of single-crystal analyses of nucleic acid structures in which increasingly larger oligonucleotides were crystallized and their three-dimensional structure solved. This was extended to large RNA molecules, and the structure of yeast phenylalanine transfer RNA was determined (Kim *et al.*, 1974; Robertus *et al.*, 1974).

In 1979 a hexanucleoside pentaphosphate with the sequence d(CpGpCpGpCpG)[2] was crystallized and solved at 0.9 Å resolution (Wang *et al.*, 1979). This double helical structure was found to exist in the form of a left-handed helix in which the two sugar phosphate chains were antiparallel, as in the A and B DNA conformations, and were held together by Watson–Crick hydrogen bonds, but the helical sense was reversed. In addition, the backbone had a curious zigzag pattern, hence the name Z-DNA. The asymmetric unit is a dinucleotide, rather than the mononucleotide as found in the A and the B conformations. We now know that there are three major families of nucleic acid double helices, the A, the B, and the Z forms. These double helices are shown in van der Waals diagrams in Fig. 1. In Figs. 1, A1, B1, and C1, the double helix is vertical and the external form of the molecule can be seen. However, in Figs. 1, A2, B2, and C2, the helical axis is tilted relative to the viewer a little over 30° so that the depth and form of the helical grooves can be more adequately visualized. This makes it possible to see the significant differences that exist in the conformation of these three forms of the double helix.

In B-DNA the double helical axis is in the center of the base pair so that the molecule has two grooves, major and minor, which differ considerably (Fig. 1, A1, and A2). The major groove has a width of almost 10 Å, and shortly after the formulation of the double helix by Watson

---

[1]A = adenine, U = uracil, G = guanine, and C = cytosine. The p represents the phosphodiester bond that links the two nucleotides.

[2]d = deoxy.

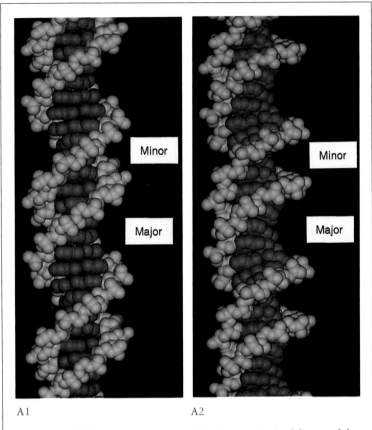

Minor

Major

Minor

Major

A1   A2

FIGURE 1. Molecular structure of the three principal forms of the DNA double helix. A space-filling model is shown in which the base pairs are colored blue and the sugar phosphate backbone is yellow. The major and minor grooves are labeled. For each model, two views are shown: one in which the model is perpendicular to the line of sight and the second in which the molecular model is tilted slightly more than 30° so that the viewer can see into the grooves in the molecule.

A. B-DNA: In A1 the molecule is vertical, and in A2 it is tipped back. Note that both grooves are accessible.

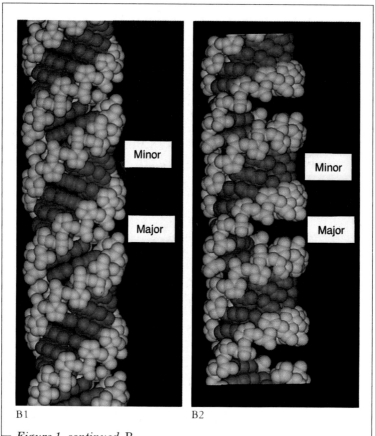

B1   B2

Figure 1, continued B.
A-DNA: In B1, the molecule is vertical, while in B2 it is tilted back. The major groove is very narrow and deep while the minor groove is broad and flat. The entrance of the major groove is quite narrow.

and Crick, it was speculated that this major groove could serve as a site for interacting with the α-helical motif of proteins, since the α-helix itself has a diameter that could be comfortably accommodated by the major groove. The minor groove was somewhat narrower and it was less clear what its role would be in protein–nucleic acid interactions.

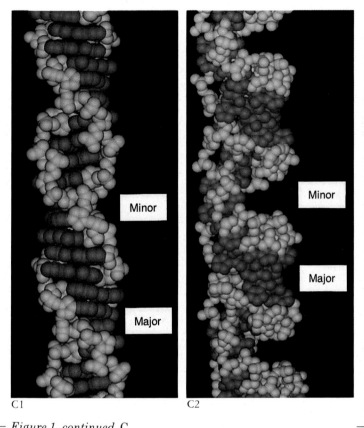

Figure 1, continued C.
Left-handed Z-DNA: The major groove is not a groove but is rather a convex surface on the outer portion of the molecule. The minor groove is rather narrow and deep, as can be seen when the molecule is tilted in C2.

In A-DNA, the helical axis no longer penetrates the base pair but rather is found in the major groove. Thus, the A-type double helix is organized somewhat as a ribbon wrapped around a central core. There is an empty cylindrical region approximately 2 Å in diameter that goes down through the center axis, and the A helix is wrapped around it. Displacement of the DNA away from the helical axis has a profound effect on the major and minor groove. The minor groove is now on the outside of the molecule where it is a broad and flat surface. The major groove is extremely deep compared to the conformation of the B helix (Fig. 1, B1, and B2). However, the entrance into the major groove is quite limited. The distance between the sugar phosphate groups of the two strands around the entrance to the major groove is less than 4 Å. This is of great importance in terms of the potential interactions of a protein with a nucleic acid in the A conformation. More specifically, for double helical RNA it means that the major groove is largely inaccessible since the side chains of proteins are unable to penetrate through the narrow opening leading to the major groove.

In the A conformation, the minor groove is fully accessible to the outside world and the major groove is largely hidden. However, in the left-handed Z-DNA conformation, just the opposite is true. There is one narrow deep groove in Z-DNA that is analogous to the minor groove. This groove is so narrow that amino acid side chains are unable to penetrate it. However, the convex exterior of the molecule is made up of the edges of the base pairs that are analogous to the major groove surface of the B-DNA conformation. Thus, in Z-DNA the major groove is accessible and the minor groove is inaccessible to amino acid side chains. This is just the opposite of what is found in A-DNA. In the B conformation, both grooves are accessible. The three major conformations of the nucleic acid double helix thus differ in the extent to which the information lying in their grooves is accessible to interaction with the amino acids of proteins.

INFORMATION CONTENT OF MAJOR AND MINOR GROOVES

An atomic-resolution structure determination of the minihelices formed by the dinucleoside monophosphates ApU (Rosenberg et al., 1973) and GpC (Day et al., 1973) determined in 1973 made it possible to address the question of information content by referring to an actual high-resolution structure determination of the double helix. The problem faced is simply this. The base pairs in a double helix are stacked so that the elements of uniqueness in them occur at the outer edges. Interactions between amino acid side chains and the outer edges of the base pairs should make it possible to discriminate one base pair from the other. There are two types of base pairs, AT[3] (or AU in RNA) and GC. Since these base pairs can occur with the purine A or G on either side, this makes a total of four types of base pairs. If we take these base pairs two at a time, there are six different possibilities. The discriminations that have to be made are illustrated in Fig. 2.

We assume that the protein uses the double-helical backbone of the nucleic acid in order to establish a frame of reference from which to probe the base pair. A pair of ribose residues in an RNA double helix is shown in Fig. 2 with two different types of base pairs superimposed, containing solid bonds or open bonds. The upper letter in Fig. 2 refers to the solid bond bases, and the lower letter refers to the open bond bases. The base pairs are drawn so that the major groove of the double helix is at the top and the minor groove at the bottom. Figure 2a shows the comparison of the A–U and the U–A base pair superimposed on each other. A methyl group would be attached to the 5 position of uracil to illustrate thymine. Figure 2b superimposes the G–C and C–G pairs. The pair U–A superimposed on G–C is also represented in Fig. 2c by rotating the pair about a vertical axis, while U–A superimposed on C–G are similarly represented in reverse order in Fig. 2d.

In our original analysis in 1976 (Seeman et al., 1976b), it was pointed out that discriminating interactions could occur by functional groups of an amino acid side chain approaching the base pair in either the major groove or the minor groove and interacting with the base pair. Potential sites for discrimination are labeled in the figure, where W stands for possible recognition sites in the major or wide groove, and S for sites in the minor or small groove. Potential recognition sites were selected if the atom is accessible for hydrogen bonding when the probe approaches it. The arrows point to the non-hydrogen atoms. Thus, in the amino group the position of the nitrogen atom is indicated, rather than the two hydrogens that are attached to it, even though recognition occurs through interactions involving the hydrogen atoms.

[3]T = thymine.

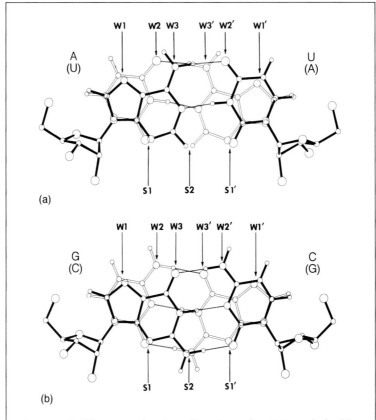

FIGURE 2. Diagram showing the stereochemistry of double-helical A·U and G·C base pairs. The geometry of the base pairs and the attached ribose residues were obtained from atomic resolution crystallographic analysis of double-helical fragments. The base pairs are superimposed upon each other, and one base pair is drawn with solid bonds and the other with open bonds. The upper letter at the side refers to the bases with solid bonds, while the lower letter in parentheses refers to the open bonds. Both bases are drawn as if attached to the same ribose

Because the molecule is an antiparallel double helix, the primed recognition sites in Fig. 2 are related to the unprimed ones of the same number by a vertical twofold rotation axis. There is no separate site S2′ as it is on the rotation axis. These recognition sites are in the same

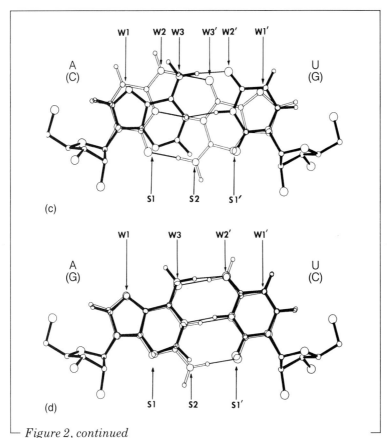

Figure 2, continued
residues in the antiparallel double helical conformation. W refers to a potential recognition site in the major or wide groove of
the double helix; S refers to sites in the minor or small groove.
The dyad axis between the two antiparallel ribose residues is
vertical in the plane of the paper. (a) through (d) represent all of
the possible base pair comparisons. (Reproduced with permission
from N. C. Seeman, J. M. Rosenberg, and A. Rich (1976b), *Proc.
Natl. Acad. Sci. USA* **73,** 804–808.)

place in all four parts of the diagram. All six possible combinations of
base pairs are shown in groups of two for ease of comparison.

An additional recognition site not shown here is the methyl group
that would be found on the 5 position of uracil if thymine were shown

rather than uracil. This hydrophobic group clearly has a unique recognition potential in the major or wide groove.

There are six potential recognition sites that are found in the major groove (Fig. 2). Ambiguities exist in discrimination because of pseudo symmetries in the base pairs (Seeman *et al.*, 1976b). The conclusion reached is that a single hydrogen-bonding probe in the major groove would probably be insufficient to discriminate uniquely certain base pairs with a high degree of fidelity. This is because small changes in the position of the protein hydrogen-bond donor or acceptor would result in confused identification. However, single hydrogen-bonding interactions could clearly discriminate a purine–pyrimidine pair from a pyrimidine–purine pair in site W1. Likewise, a single probe in sites W2 or W3 could differentiate a base pair containing either an adenine or cytosine on the left-hand side of the pair from those base pairs containing either uracil (thymine) or guanine there. Thus, a single hydrogen-bonding probe in the wide groove is able to resolve two different types of ambiguity, depending upon whether the probe involves site W1 or, alternatively, sites W2 or W3.

In the minor groove we see a very different geometric and electrostatic environment. Figure 2 shows that there are only three sites on this side of the base pair that contain functional groups. S1 and S1′ are symmetrically positioned by the vertical dyad axis that relates the sugar residues of the antiparallel chains, while S2 is located directly on this dyad axis. Analysis of minor groove interactions leads to ambiguity (Seeman *et al.*, 1976b). Recognition in the minor groove is likely to be relatively insensitive to base pair reversals, as shown in Fig. 2a for A–U(T) versus U(T)–A. Nonetheless, an interaction in site S2 will be capable of making the discriminations seen in Fig. 2c or 2d. The situation for discrimination due to single hydrogen-bonding interactions in the minor groove, like those described in the wide groove, leads to further ambiguities.

There are fundamental limitations in the discrimination of individual base pairs in either groove by a single hydrogen-bonding interaction because of the difficulty of fixing the precise position of the hydrogen-bond donor or acceptor. While small movements of amino acid side chains are likely to occur, the situation may be quite different with hydrogen-bond donors or acceptors involving the polypeptide backbones, as these are more likely to be constrained in space.

Pairs of Hydrogen Bonds Are
Better Than Single Ones

An analogy can be made between the way polynucleotides are re-
sponsive to base sequences in double-helical nucleic acids. For example, the polynucleotide double helix $(rA:rU)$[4] can add with great
specificity a third strand of polyuridylic acid that interacts with the
double helix using two hydrogen-bonding recognition sites (Felsenfeld
et al., 1957). A large number of highly specific polynucleotide interactions occur, all of which have the characteristic property of utilizing
two hydrogen bonds as the basis for specificity (Davies, 1967). These
analogies were used to suggest a system in which amino acid side
chains form similar specific pairs of hydrogen bonds with base pairs in
the double helix. Examples are shown for the major groove in Fig. 3
and for the minor groove in Fig. 4. Figure 3 shows the type of hydrogen
bonding in which the uracil residue interacts specifically with the A–U
pair in the triple-stranded complex (Felsenfeld *et al.*, 1957). Sites W1′
and W3′ of adenine are used to form two specific hydrogen bonds with
uracil O4 and N3–H. This may be regarded as an analogue of the proposed interaction in which amide side chains of amino acids asparagine or glutamine could form a similar pair of hydrogen bonds. In
Fig. 3b a hydrogen bonding interaction is shown between guanine
residues using sites W1′ and W3′ of the guanine in a C–G base pair.
This arrangement is found in the three-dimensional structure of yeast
tRNA[Phe] (Kim *et al.*, 1974; Robertus *et al.*, 1974).[5] The guanine amino
group N2 and N1–H serve as hydrogen bond donors to the guanine
atoms O6 and N7 at sites W3′ and W1′. This was suggested as an
analogue of the proposed interaction in which the NH_2 groups of the
guanidium side chain of arginine are shown forming hydrogen bonds
to the same sites. In Fig. 3b we have arbitrarily used the two guanidi-
nium amino groups in this interaction, although it is clear that one
amino and one imino group could also be used as a pair of donors.

Figure 4 shows a way in which sites S1′ and S2 might be used for
discriminating the C–G base pair in the minor groove. On the left we
see how guanine interacts with the minor groove of the G–C pair as

---

[4] r = ribo–.
[5] Phe = phenylalanine; tRNA = transfer RNA.

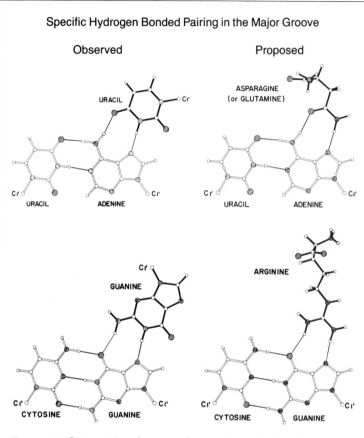

FIGURE 3. Interactions between base pairs and other bases (observed) or amino acid side chains (proposed). (a) The uracil binding to the U·A pair is seen in polynucleotides as well as in single crystals of adenine and uracil derivatives. The guanine binding to the C·G pair is seen in yeast phenylalanine tRNA (Kim *et al.,* 1974). Oxygen atoms have diagonal shading, while nitrogen atoms are stippled. (Reproduced with permission from N. C. Seeman, J. M. Rosenberg, and A. Rich (1976b), *Proc. Natl. Acad. Sci. USA* **73,** 804–808.)

seen in the crystal structure of 9-ethylguanine and 1-methylcytosine (O'Brien, 1967). In this crystal structure the amino group N2 of guanine acts as a donor to site S1′, while N3 acts as an acceptor from site S2. An analogue to this may be seen in the proposed interaction in

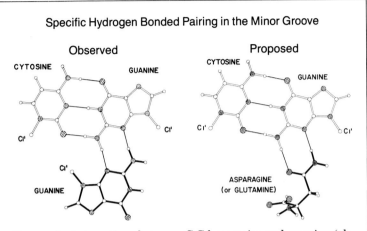

FIGURE 4. Interactions between C·G base pairs and guanine (observed) or asparagine (proposed). The guanine interaction with the C·G pair is seen in the structure of the 9-ethyl guanine·1-methyl cytosine complex (O'Brien, 1967). (Reproduced with permission from N. C. Seeman, J. M. Rosenberg, and A. Rich (1976b), *Proc. Natl. Acad. Sci. USA* **73**, 804–808.)

which the amide of asparagine or glutamine forms a pair of hydrogen bonds with the same two sites of the guanine in the C–G base pair.

Thus, interactions with amino acid side chains using pairs of hydrogen bonds are able to differentiate unambiguously A–U(T) or G–C pairs in the major groove of the double helix, and possibly G–C pairs in the minor groove. However, there is not an analogous similar system for differentiating the A–U base pair from the U–A base pair by an interaction in the minor groove.

The principle of using two hydrogen bonds to specify a base pair is a reasonable but not necessarily unique recognition mechanism. The initial description only considered recognition by single amino acid side chains (Seeman *et al.*, 1976b). The analysis could have been extended to using pairs of amino acid side chains to effect recognition. As noted below, several examples of this are found in protein–nucleic acid complexes.

In 1976 when this attempt was made to deduce rules associated with protein–nucleic acid recognition, there were no examples of crystal structures that had been solved containing a protein bound in a sequence specific manner to a nucleic acid fragment. At the present time, there have been several structures of this type solved, and what is apparent in the following segments is that the mechanism of identifying nucleic acid sequences is more complex than could have been anticipated in 1976. For example, we now know that the architecture of the double helix is often changed, with changes in helical twist, buckling, or propeller orientation of base pairs. This will, of course, affect recognition systems. Nonetheless, some of the recognition mechanisms that were suggested then have indeed been found in a variety of protein–nucleic acid crystal structures.

PROTEIN STRUCTURE

One of the great contributions made by Linus Pauling over forty years ago was the enunciation of the structural motifs found in proteins. These were deduced by Pauling through a study of the detailed structural geometry of simple peptides. The description of the α helix and the β sheet with both parallel and antiparallel conformations anticipated the discovery of these major structural motifs in a variety of proteins through x-ray crystallographic analysis. Not surprisingly, these motifs are used extensively in recognition of nucleic acids by proteins. Table I outlines the recognition motifs that are found in a number of protein–DNA complexes. This list is of necessity incomplete, since it only contains structures whose three-dimensional structures have been determined. A number of other sequence-specific DNA binding proteins have been identified, but their mode of recognition of DNA is yet to be determined.

THE HELIX-TURN-HELIX MOTIF AND DNA RECOGNITION

Proteins that regulate the transcription of DNA bind in a sequence-specific manner and generally contain discrete DNA binding domains in their structure. These binding domains are frequently less than 100 amino acids in size. The first of these to be studied extensively were DNA recognition domains obtained from bacterial or bacteriophage regulatory molecules. These regulatory molecules are frequently repressive in action and prevent RNA polymerase from transcribing

TABLE I

CRYSTAL STRUCTURES OF PROTEIN–DNA RECOGNITION MOTIFS

Motif	Resolution (Å)	Reference
A. Helix Turn Helix–DNA Complex		
434 repressor fragment–DNA	4.5–3.5	Anderson *et al.* (1987)
434 repressor fragment–DNA	2.5	Aggarwal *et al.* (1988)
λ repressor fragment–DNA	2.6	Jordan and Pabo (1988)
434 cro repressor–DNA	4.0	Wolberger *et al.* (1988)
λ cro repressor–DNA	3.9	Brennan *et al.* (1990)
E. coli CAP–cAMP–DNA	3.0	Schultz *et al.* (1990)
		Steitz (1990)
E. coli Trp repressor–DNA	2.4	Otwinowski *et al.* (1988)
Engrailed homeodomain fragment–DNA	2.8	Kissinger *et al.* (1990)
B. Zinc Fingers–DNA Complex		
Zif 268 fragment–DNA	2.1	Pavletich and Pabo (1991)
Glucocorticoid receptor–DNA	2.9	Luisi *et al.* (1991)
C. Endonuclease RI–DNA Complex		
E. coli Eco RI–DNA	3.0	McClarin *et al.* (1986)
E. coli Eco RI–DNA	2.8	Kim *et al.* (1986)
D. β Sheet DNA Binding Proteins		
E. coli Met Repressor	2.8	Rafferty *et al.* (1989)

DNA, or sometimes they activate and facilitate the initiation of transcription. Many three-dimensional x-ray structures have been determined from proteins belonging to the bacteriophage λ and other related phages such as 434 or P22. These phages code for two regulatory proteins, namely repressor and cro. Cro is also a repressor protein, and both of them act to regulate transcription of DNA and form the components of an important genetic switch. The bacteriophage can be incorporated into the bacterial cell genome and the bacterial cell divides without expressing bacteriophage proteins. The bacteriophage lies dormant during normal cell growth. However, the bacteriophage can be lytic for the host bacterial cell. Brief exposure to ultraviolet light switches on the phage genes, which then produce new phages and eventually lyse the bacterial cell. The mechanism for this genetic on–off switch has been elucidated by a combination of x-ray structural studies and genetic studies that give us a detailed understanding of the process.

A small region of the phage genome contains the genetic components of the on–off switch. This region has two structural genes coding for the two regulatory proteins, cro and repressor. They are transcribed in opposite directions, and between them sits a regulatory or promoter region where the polymerases bind. When the polymerase acts in one direction, synthesis of cro is initiated together with the early lytic genes. This results in the lysis of the cell. However, when the polymerase is bound to the other promoter, synthesis of repressor is turned on, and cro and the lytic genes are not expressed. The bacterial cell then survives and continues normal replication. The decision to lyse or not to lyse depends upon which of the two promoters in the operator region is able to bind to polymerase. That depends on the binding of cro and repressor protein to three binding or operator sites situated in the regulatory region. The binding sites for the cro promoter and for the repressor promoter overlap somewhat. The functioning of the lytic switch critically depends upon the relative binding of cro or repressor molecules to the three different binding sites in the operator. In phage λ, these three binding sites are similar but not identical. They are also partly palindromic at their ends. The palindromic arrangement is important because it means that the two halves of each binding site are related by approximate twofold symmetry axes. Both cro and repressor bind as dimers to all three sites, but the affinity of the binding varies somewhat. Because the protein is dimerized, it means it is interacting with both halves of the palindromic sequence.

Cro is a small protein that forms stable dimers in solution. It has 66 amino acid residues, and its structure was solved by Matthews and his colleagues in 1981 (Anderson et al., 1981). They found that it folded into three α helices, and the structure also contained three strands of antiparallel β sheet that was used in dimerizing the protein. The α helices were found in a unique arrangement with one helix protruding from the molecule. It was separated by a β turn to another helix. This arrangement of a helix-turn-helix is also found in λ repressor and in the Escherichia coli (E. coli) catabolite-activating protein (CAP) (McKay and Steitz, 1981).

In 1981, Matthews and his colleagues realized that the dimer structure of cro had an important structural consequence relative to its mode of interacting with DNA. The protruding helices of the helix-turn-helix motif of the two monomers were 34 Å apart, which corresponds almost exactly to the separation between two major grooves of DNA along one turn of the B-DNA double helix. They were able to build a model of a possible cro–DNA complex in which the two recognition helices of the cro dimer fitted into the major groove of a piece of DNA (Anderson *et al.*, 1981). The orientation of the protruding or recognition α helices follows the orientation of the major groove, and amino acids from these α helices are in a position to make contact with the edges of base pairs in the major groove where they could be involved in recognizing specific nucleotide sequences. The centers of the palindromic sequences in the DNA binding regions are separated by about 10 base pairs or 34 Å. Thus, if the recognition helix of one part of the dimer is interacting with one end of the palindrome, then the other is also interacting with the same sequence at the other end.

This arrangement of molecules implies that part of the recognition helix determines the specificity of the binding of the entire protein. The consequences of this model were tested experimentally in genetic studies in which the amino acid sequence of the repressor from phage 434 had its recognition α helix replaced with that of the corresponding α helix from cro of phage 434. The redesigned repressor then acquired the differential binding properties of the 434 cro protein. Similar experiments were done changing selected amino acids in the recognition α helix of 434 repressor to those that are found in the repressor of phage P22 (Wharton and Ptashne, 1985). These amino acids were selected because they were on the outer part of the α helix that presumably interacts with the DNA. The parental 434 repressor had no affinity for the P22 operator *in vivo,* but the redesigned 434 repressor worked with P22 operators and not with 434 operators *in vivo.* Thus, the genetic experiments clearly showed that the proposed structural model for the interaction of the recognition helix with the DNA was essentially correct.

As listed in Table I, both repressor and cro DNA cocrystals have had their three-dimensional structure solved for both λ bacteriophage and 434. This has made it possible to define in some detail the interaction of the amino acid side chains of the protein with the DNA. What emerges from these studies is the fact that a large number of recognition interactions are seen, including some of those that were postulated by the analysis of DNA sequence-specific binding carried out in 1976 (Fig. 3).

The arrangement of the helix-turn-helix interactions found by Jordan and Pabo in the λ repressor–DNA complex is shown in Fig. 5. Here it can be seen that the recognition helix 3 is more or less lying in the major groove of the DNA, while helix 2 of the helix-turn-helix motif is organized across the groove so that it interacts with the DNA backbone at either end. A number of interactions are found between amino acid side chains and base pairs, three of which are illustrated in the figure. These show the interaction of a serine residue with the N7 of guanine in base pair 4. It also shows the interaction of an asparagine and lysine side chain with N7 and O6 of guanine in base pair 2. Thus, two amino acids cooperate to form a pair of recognition hydrogen bonds. Finally, base pair 6 has recognition of an adenine by a glutamine bonding to N6 and N7 in a manner similar to that described in Fig. 3. It is interesting that the glutamine 44 in base pair 6 is further stabilized in its interaction by hydrogen bonding to another glutamine 33 that interacts with the phosphate group of the backbone.

The complexes of 434 cro and repressor with their DNA operators were studied by Harrison and his colleagues (Aggarwal *et al.*, 1988; Wolberger *et al.*, 1988), and they showed that the binding domains are very similar to each other. 434 cro has 71 amino acids, and these have a 48% sequence identity to the 69 residues that form the *N*-terminal DNA domain of 434 repressor that binds to the DNA. Both of them dimerize and have helix-turn-helix motifs at either end of the dimer that protrude in such a manner that they can interact with the DNA. Even though these 434 cro and repressor fragments are monomers in solution, they were found dimerized in the structure that was solved. Cro is a small molecule and the repressor is quite large. The structure solved had only the DNA binding domain of the repressor, and it is likely that the other segment of the repressor contains additional dimerizing interactions. In these complexes, the DNA is in the B form, but there are significant distortions in it. Thus, there are changes in the local twist so that it is overwound in its center and unwound at its ends. In addition, the helical axis was bent somewhat in a manner that narrowed the minor groove in the center and widened it at its ends. These conformational changes in the DNA are due to the interactions with the protein. It is important to note that the DNA conformations differ in the complex with cro compared to the repressor complex. These differences are probably due to the differences in the identity of

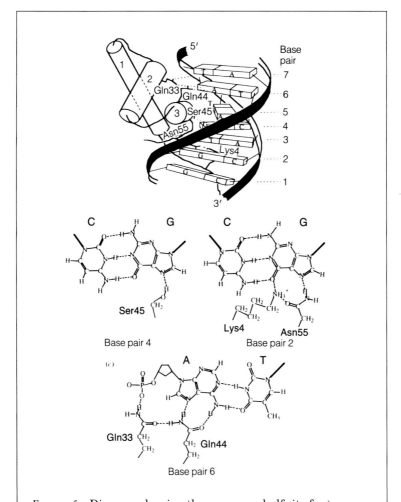

FIGURE 5. Diagram showing the consensus half site for λ repressor bound to a DNA fragment. The α helix is drawn as a cylinder, and the helix-turn-helix motif is closest to the DNA. The recognition helix 3 sits in the major groove of the DNA where its side chains can interact with the bases. At the bottom there are diagrams illustrating how the side chains hydrogen-bond to base pairs 2, 4, and 6 in the consensus half site. The heavy black lines show the connection to the sugar phosphate backbone. It can be seen that gln 33 is hydrogen-bonding both to the phosphate group of adenosine and the gln 44 side chain. (Adapted from S. R. Jordan and C. O. Pabo, *Science* **242,** 893–899, Copyright © 1988 by the AAAS.)

the amino acids found in the recognition helix. Furthermore, it is likely that these conformational changes are important for the differential repressor binding. The repressors bind to the three different palindromic sequences with varying affinities that are important in the mechanism of the genetic switch. In the binding of 434 repressor to the operator DNA (Aggarwal *et al.*, 1988), there are interesting interactions of the side chains of glutamine 28 and glutamine 29 with the base pairs. Glutamine (Gln) 28 forms two hydrogen bonds to N6 and N7 of adenine of an AT base pair in a manner similar to that shown in Fig. 3. However, Gln 29 forms a bifurcated hydrogen bond to O6 and N7 of guanine in an adjacent GC base pair using the amino group of Gln. Its conformation is such that the side chain also forms a hydrophobic pocket that is interacting with the thymine of an adjacent AT base pair. The methyl groups of the side chains of a threonine 27 and glutamine 29 form a hydrophobic pocket that contains the methyl group of the thymine. In this one region, it illustrates a variety of different interactions that are found between 434 repressor fragment and its operator.

The protein–DNA contacts in this complex are responsible for an overwinding of DNA in the central region that is associated with the narrowing of the minor groove. It is generally recognized that AT and TA base pairs are found in regions in which the minor groove can be made narrower. This cannot occur in sequences that have GC or CG base pairs because of the presence of an amino group in the 2 position of guanine. Thus, if sequences in the centers of the palindromic operators have AT base pairs changed to GC base pairs, the affinity of the repressor fragment is decreased significantly. The base sequence is seen to moderate the extent to which local DNA structure can be distorted, which in turn alters the affinity of the repressor for DNA. Thus, the ability to undergo a conformational change in the DNA is an important component of some protein–nucleic acid recognition systems. Several examples of conformational changes have been observed in other studies as well.

A number of the protein–DNA complexes have base pair recognition through single amino acids that form two hydrogen bonds, such as those described in Fig. 3. These are listed in Table II for a variety of proteins, including several that have the helix-turn-helix motif. It can

TABLE II

Protein–DNA Cocrystals Showing Single Amino Acid–Nucleotide Recognition
Using Hydrogen Bond Pairs

Structure	Amino Acid–Base Interaction	Reference
Eco RI–DNA	Arg 200 to guanine	McClarin *et al. Science* **234,** 1526 (1986)
λ repressor–DNA	Gln 44 to adenine	Jordan and Pabo, *Science* **242,** 893 (1988)
434 repressor–DNA	Gln 28 to adenine	Aggarwal *et al., Science* **242,** 899 (1988)
434 cro repressor–DNA	Gln 28 to adenine	Wolberger *et al., Nature* **335,** 789 (1988)
Trp repressor–DNA	Arg 69 to guanine	Otwinowski *et al., Nature* **335,** 321 (1988)
Homeodomain–DNA	Asn 51 to guanine	Kissinger *et al., Cell* **63,** 579 (1990)
λ cro repressor–DNA	Gln 27 to adenine Arg 38 to guanine	Brennan *et al., Proc. Natl. Acad. Sci.* **87,** 8165 (1990)
Mouse zinc finger protein–DNA	Arg 18 to guanine 10 Arg 24 to guanine 8 Arg 46 to guanine 7 Arg 74 to guanine 4 Arg 80 to guanine 2	Pavletich and Pabo, *Science* **252,** 809–816 (1991)
Glucocorticoid receptor–DNA	Arg 466 to guanine 4	Luisi *et al., Nature* **252,** 497 (1991)

be seen that this recognition system is used frequently in these interactions, but it is by no means a recognition code in that it is only one
of several types of interactions that are found in the crystal structures.
However, as will be seen in what follows, this type of interaction does
occur in a manner that suggests a recognition code in the interaction
of a zinc finger protein with a DNA sequence.

The variety of interactions found between proteins and DNA can
be illustrated in the λ repressor–DNA structure by Jordan and Pabo
(1988). Figure 5 shows some of the interactions of this helix-turn-helix
motif, and Table III gives a more complete listing of not only the major
groove interactions, but also the interactions with the DNA backbone.

TABLE III

λ REPRESSOR–DNA INTERACTIONS[a]

Major Groove Interactions	DNA Backbone Interactions
Hydrogen bonding groups	P_A
Adenine 2	O-1P . . . Gln-33 N
N-6 . . . Gln-44 OE1	O-2P . . . Tyr-22 OH
N-7 . . . Gln-44 NE2	O-2P . . . Lys-26 NZ
Guanine 4	P_B
N-7 . . . Ser-45 OG	O-1P and O-2P . . . Lys-19 NZ
Guanine 6	O-1P . . . Gln-33 NG2
N-7 . . . Asn-55 ND2	O-1P . . . Asn-52 ND2
O-6 . . . Lys-4 NZ	P_C
Guanine 8	O-1P . . . Gly-43 N
N-7 . . . Thr-2 OG1	P_D
Hydrophobic interactions	O-1P and O-2P . . . Asn-61 ND2
Thymine 1	P_E
C-5 M . . . Gln-44 CG	⌈ O-1P and O-2P . . . Asn-58 ND2
Thymine 3	(in consensus half-site)
C-5 M . . . Ile-54 CD1	O-1P . . . Ala-56 N
Gly-48 main chain	(in nonconsensus half-site)
Ser-45 CB and main chain	
Thymine 5	
C-5 M . . . Gly-46 CA	
Ser-45 CA and CB	

[a] Adapted from Sauer *et al.* (1990).

The major groove interactions include the three that are shown in Fig. 5, as well as hydrophobic interactions involving the methyl group on three different thymine residues. The hydrophobic interactions used for defining the position of thymine residues are an important component in all of these recognition systems. The interactions between the DNA backbone and the proteins listed in Table III are important in maintaining the conformation of the DNA in the complex. The structure of this complex was important in illustrating the manner in which pairs of amino acids can cooperate to enhance the specificity of recognition. Two examples of this are shown in Fig. 5 where the side chains from two amino acids are used to identify both adenine and guanine residues.

In the λ repressor operator complex, there are no significant distortions in the DNA conformation, in contrast to those seen in the interactions of 434 repressor and cro with DNA. Thus, this suggests that conformational changes may or may not be found in DNA when it interacts with recognition proteins. An example in which there is significant bending of the DNA is found in the interaction of the catabolite activating protein (CAP) bound to its DNA (Steitz, 1990). Here the DNA fragment induces a sharp bend in the DNA that may be an important component in activating the DNA for transcription.

The three-dimensional structure of the tryptophan repressor has been solved, both with and without L-tryptophan complexed to it (Schevitz et al., 1985; Lawson et al., 1988). Tryptophan is used to activate the repressor for binding, and the mechanism for that has been worked out in some detail by Sigler and his colleagues. Binding of tryptophan produces a conformational change in the dimeric protein, such that the recognition helices of the trp helix-turn-helix motif are changed in orientation so that they are now capable of interacting with the major groove of DNA in the presence of the tryptophan affector. This is called an allosteric change, in which the small molecule influences the conformation of the large molecule and in turn activates it. The trp repressor thus has a simple negative feedback loop in which a decreased concentration of tryptophan in the cell inactivates the repressor by losing its allosteric activator, thereby allowing tryptophan biosynthesis to proceed.

The structure of the tryptophan (Trp) repressor bound to DNA has been worked out in great detail (Otwinowski et al., 1988). The recognition in this complex appears to be mediated by a variety of bound water molecules that are interposed between the recognition helix of the helix-turn-helix motif and the major groove of the DNA. This structure has provoked a great deal of discussion regarding the role of buried water molecules in DNA recognition. Some investigators feel that the structure solved in the trp repressor–DNA complex is actually that of the nonspecific complex rather than the specific complex because of the presence of the water molecules. However, more information will be required before we can readily assess the role of the buried water molecules in the trp repressor–DNA complex in molecular recognition.

The prokaryotic proteins with helix-turn-helix motifs generally occur as dimers when they recognize DNA. However, an important eukaryotic protein is the so-called homeodomain, which binds to DNA in a sequence-specific manner and uses a helix-turn-helix motif as a monomer rather than as a dimer. The homeobox is a DNA sequence of 180 base pairs that codes for homeodomain proteins. The homeobox was first discovered in the fruit fly *Drosophila*. Homeotic transformations are those in which mutations cause unusual disturbances in the fly's developmental body plan. Homeoboxes have been found in many different species, including vertebrates and invertebrates, and it is believed that they play a key role in embryological development. The homeodomain proteins, about 60 amino acids long, probably function largely as regulatory transcription factors.

The engrailed protein is one of the members of this family, and Pabo and his colleagues have solved the structure of a 61-amino-acid peptide fragment from the engrailed protein bound to its DNA binding site (Kissinger *et al.*, 1990). In Fig. 6 it can be seen that the recognition helix 3 lies in the major groove of the DNA where its side chains can interact with the base pairs. The other helix 3 spans the sugar phosphate chains of the two DNA strands. Two of the interactions of side chains with base pairs are shown, including a hydrophobic interaction of an isoleucine side chain with the methyl group of thymine of an AT base pair, and the interaction of an asparagine residue with adenine in a manner similar to that which was described in Fig. 3. In this complex, the DNA duplex is a relatively straight segment of DNA without any significant distortions. However, there is some widening noted in the major groove where helix 3 binds. This binding is associated with a tilt of the base pairs that may be related to the change of the groove width.

Although there is a general overall similarity between the helix-turn-helix motifs in the eukaryotic and in the prokaryotic systems, there are nonetheless some small but significant differences in the detailed positioning of helix 2 and helix 3 relative to the DNA compared with that seen in the prokaryotic helix-turn-helix proteins. Another difference is that the helices in the homeodomain are longer than the corresponding helices of the prokaryotic repressors. In the longer recognition helix, the residues contacting the bases are near the center of this extended helix.

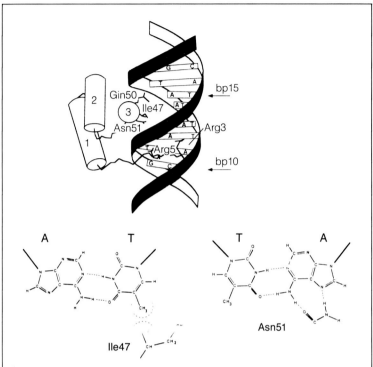

FIGURE 6. Diagram showing the interactions of the engrailed homeodomain with double stranded DNA. The recognition helix 3 of the helix-turn-helix motif is engaged in the major groove of the DNA where its side chains interact with the bases. Note that there are also some minor-groove contacts. Figures at the bottom illustrate the base contacts made by two of the amino acids. (Adapted from Kissinger *et al.* (1990), *Cell* **63,** 579–590, with permission.)

ZINC FINGERS ARE AN IMPORTANT PROTEIN MOTIF THAT INTERACTS WITH DNA

The zinc finger is one of the major structural motifs found in proteins that bind to DNA. These are eukaryotic proteins that frequently act as transcription factors. The fingers were first identified through a sequence motif of the form $X–Cys–X_{2-4}–Cys–X_{12}–His–X_{3-4}–His–X_4$,[6]

[6] Cys = cystine; His = histidine.

where X is any amino acid. Hundreds of similar finger sequences have been identified. This is an important class of zinc-containing proteins. There are other proteins that contain zinc with other binding motifs. For example, the glucocorticoid receptor has four cystenes that bind to a zinc atom. Its structure will be discussed below. The typical zinc finger, however, is characterized by the presence of two cystine residues and two histidine residues that complex two zinc atoms in a tetrahedral fashion. Nuclear magnetic resonance studies have shown that this type of zinc finger contains an antiparallel β ribbon and an α helix. The two invariant histidines are near the COOH end of the α helix, while the two invariant cystines are near the turn of the β ribbon. These coordinate a central zinc ion and form a compact domain. The zinc finger may be looked upon as a miniglobular protein with a hydrophobic core and polar side chains on the surface. Thus, nonpolar residues are found in the region between the antiparallel β sheet and the α helix that, together with the zinc coordination, stabilizes this conformation.

Recently a zinc finger–DNA crystal structure has been solved by Pavletich and Pabo (1991) that involved a segment of a mouse regulatory protein. They have cloned the DNA binding domain with three zinc fingers and crystallized this peptide (Zif 268) with a consensus DNA binding sequence containing 10 base pairs. Zinc fingers are frequently found in variable numbers. Some proteins have one or two zinc fingers, while others may have large numbers up to over 30 segments. For some time there has been considerable debate concerning the mode of interaction of successive zinc fingers that are connected by linker peptide to DNA. The Zif 268–DNA complex resolves this issue and shows that the three zinc fingers are resting in the major groove of the DNA double helix in a more or less continuous fashion, one after another.

The three zinc fingers are arranged in a semicircular or C-shaped structure that fits tightly into the major groove of B-DNA (Fig. 7). The cocrystal shows that the α helix of each zinc finger fits directly into the major groove, and that residues from the N-terminal portion of the α helix contact the base pairs in the major groove. Each of the three zinc fingers in the complex has its α helix binding to the DNA duplex in a similar fashion. In this way, each finger makes primary contacts with

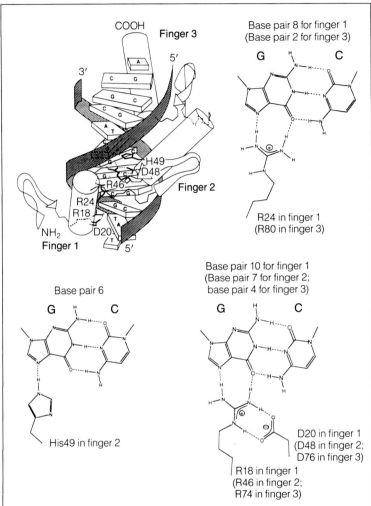

FIGURE 7. Diagram showing three zinc fingers from the Zif 268 protein interacting with DNA in the consensus sequence. The three zinc fingers, consisting of a cylindrical α helix and a β sheet, have identical interactions with the successive triplets of bases in the major groove of the double helix. The diagrams at upper right and bottom illustrate the types of hydrogen-bonding interactions that are found in the different zinc fingers. Arginine binding to guanine residues is found in five of the six hydrogen-bonding interactions. (From N. P. Pavletich and C. O. Pabo, *Science* **252**, 809–816, Copyright © 1991 by the AAAS.)

a three-base-pair subset. Thus, each finger is related by a rotation of 3 × 32° and a translation of about 10 Å (3 × 3.3 Å) along the DNA axis. Unlike the recognition helix of the helix-turn-helix motif, the α helix in the zinc finger is tipped at an angle away from the major groove. The β sheet is on the back of the helix away from the base pairs and is shifted towards one side of the major groove. The two strands of the β sheet have different roles. The first β strand does not make contact with the DNA, whereas the second β strand is in contact with the sugar phosphate backbone along one strand of DNA.

The zinc finger peptide makes 11 hydrogen bonds with the bases in the major groove. Six amino acid side chains interact with the base pairs, two from each zinc finger. However, five of the six residues that are binding to the nucleic acids are arginine residues. One of these arginines immediately precedes the α helix in each of the three fingers, and it also includes the residues from either the second, the third, or the sixth residues in the α helix. All of these form hydrogen bonds with the G-rich strand of the consensus binding site. Figure 7 shows the interactions of the various arginine residues and one histidine residue from the three fingers. It can be seen that five of the six interactions are those postulated at an earlier time in an analysis of protein–nucleic acid recognition (Fig. 3).

Each zinc finger interacts with a subset of three bases. All of them interact with the first base in the subset. However, two of them interact with the third base of the subset and one with the second base. The detailed interactions are listed in Table II and are also shown in Fig. 7. Some of the arginine residues binding to guanine are also stabilized by aspartic acids that occur as the second residue in the α helices. It is likely that this side-chain interaction helps to stabilize the binding of the arginine to the guanine residue, although it is not found in all of the fingers. The recognition system is relatively simple. The residue immediately preceding the α helix contacts the third base on the primary strand of the substrate at the 5′ end. The third residue on the α helix can contact the second base on the primary strand, and the sixth residue can contact the first base. In this structure, each zinc finger came in contact with only two bases of each three-base subset. It is not known whether it would be possible to have other zinc fingers in which all three bases are recognized.

The DNA is essentially in the form of a B-type helix with small distortions. There are small changes in base-pair twist going from one site to the next, although the overall conformation is similar to normal B-DNA. Each of the three fingers binds in a similar orientation and has similar contacts with the three base pairs of the DNA. In a formal sense, the relationship of one zinc finger to the next is in the form of a translational rotation or a screw operation that tracks the inside of the major groove of the double helix.

Even though the zinc finger uses the α helix for recognition, it has several unique features that differentiate it from the helix-turn-helix interactions. First of all, the zinc finger complex is formed out of modular units that can be repeated a large number of times. This entails the possibility of recognizing very long stretches of DNA by simply having a larger number of zinc fingers. As mentioned, some proteins have very large numbers of zinc fingers and may actually use this. The second characteristic is that the contacts seem to be largely with one strand only of DNA, in this particular case with the purine-rich or guanine-rich strand. The recognition depends largely on interaction with the bases, and there are fewer hydrogen bonds with the DNA backbone than are seen in the other structures.

Even though studies of other protein–DNA complexes appear not to have a recognition code, the zinc finger complex appears to have a recognition code which is largely based on the arginine–guanine contacts, at least for the Zif complex. It remains to be seen whether these types of contacts will be found more specifically when the structures of more zinc fingers have been done. Solution of this structure will make it possible to synthesize new zinc finger binding domains with different nucleotide binding specificities. Thus, it will be possible to explore the full gamut of interactions found in this widely used recognition motif in eukaryotic systems.

THE GLUCOCORTICOID RECEPTOR ALSO CONTAINS ZINC IONS

The glucocorticoid receptor has the property of binding a hormone, such as estrogen or another steroid, and is then translocated from the cytoplasm to the nucleus, where it binds to specific DNA sequences, called *glucocorticoid response elements* (GREs). The binding affects transcription of the genes. A large number of these exist; they include receptors for steroid hormones, retinoids, vitamin D, thyroid hormones, and others. Members of this superfamily have a common amino acid sequence organization with discrete domains that are used for binding DNA as well as zinc. All of these nuclear receptors are characterized by a pattern containing eight cystenes and, in the glucocorticoid receptor, these cystenes coordinate two zinc ions in a tetrahedral manner. The structure of the glucocorticoid receptor bound to DNA has been determined recently by Luisi *et al.* (1991). Unlike the typical zinc fingers, the glucocorticoid receptor forms a distinct globular binding domain and does not occur in a long series of modular units, as is often found in the typical zinc finger DNA binding.

The structure of the glucocorticoid receptor bound to DNA reveals that the receptor dimerizes onto a DNA molecule that contains two repeats of the glucocorticoid response element sequence, each with the major groove facing in the same direction. Each of the proteins forms a compact globular structure in which the two zinc ions serve to nucleate the formation of a conformation in which an α helix is positioned in the major groove of B–DNA and thereby has sequence-specific binding. The glucocorticoid receptor conformation may be looked upon as a conformation similar in some respects to the helix-turn-helix conformation, except that zinc ions are used in maintaining the stable fold of the protein rather than the helix interactions found in the helix-turn-helix system. A number of interactions are found between the glucocorticoid receptor and the DNA. However, three of them are interactions with bases that are important for determining sequence specificity. One of the most important of these is an arginine 466 that binds to guanine 4 using the system of arginine–guanine interactions, which has been described above using two hydrogen bonds. Another hydrophobic interaction involves valine 462 interacting with the methyl group of thymine 5 while a lysine 461 forms a single hydrogen bond to N7 of guanine 7 as well as to a water molecule, which in turn binds to O6 of guanine and O4 of thymine in an adjacent base pair. If arginine 466 is replaced by lysine or glycine, the protein no longer functions *in vivo*. Arginine 466 and lysine 461 are absolutely conserved in the superfamily of nuclear receptors; their targeted bases, guanine 4 and guanine 7, also occur consistently in all the known sequences of the hormone response elements (Luisi *et al.*, 1991).

The major difference between the zinc-containing glucocorticoid response element and the traditional zinc finger is the fact that the latter conformation is stabilized individually by an extensive hydrophobic core as well as by the zinc ion. Furthermore, it assumes this conformation independent of the presence or absence of DNA. In contrast, experiments with the glucocorticoid receptor show that it only condenses as a dimer in the presence of the DNA. The dimerization is stabilized both by the DNA as well as by contacts between the protein.

The arginine–guanine interaction, which played so predominant a part in five of the six interactions seen in the three modules of the traditional zinc finger structure, also plays a role in interactions with the glucocorticoid response element. However, in this case, only one of the three interactions that are sequence-determining involves the arginine–guanine interaction.

ECO RI ENDONUCLEASE BINDING TO DNA

Restriction endonucleases are very important tools in molecular biology since they cleave DNA molecules at specific sequences. One of the widely used enzymes was obtained from *E. coli* and is called Eco RI endonuclease. It cleaves DNA at a specific double-stranded sequence (d(GAATTC)). Eco RI contains 276 amino acids, and it has been crystallized with a fragment of DNA containing 13 base pairs. The solved structure revealed a complex interaction between a globular protein and a DNA double helix (McClarin *et al.*, 1986; Kim *et al.*, 1986). The DNA recognition motif consists of a parallel bundle of four α helices penetrating the major groove of the DNA. There, amino acids at the end of the α helix interact with the bases in the major groove. Although α helices are employed, this motif differs from the interaction seen both in the helix-turn-helix proteins and the zinc fingers. In this case, a cluster of very long α-helical segments interact with the DNA at their ends. In addition, a segment of extended polypeptide chain runs along the major groove of the DNA roughly parallel to the DNA backbone. This is anchored at one end by one of the recognition helices, and it has several contacts with bases. Among the interactions that are described is one involving arginine 200 binding to guanine in a manner similar to that described in Fig. 3.

We do not know whether this structure is likely to be a general structure for the recognition of DNA by restriction endonucleases. However, one of the interesting projects arising from solution of this protein–DNA complex is the possibility of modifying side chains to alter recognition modes so that one might be able to make restriction enzymes with altered cleavage specificities using the Eco RI framework for carrying this out. Further work will be necessary before we know whether this is a general recognition motif for other enzymes as well. However, it is important to emphasize that the mode of interaction is quite distinct from that seen in any other protein–nucleic acid cocrystal.

β SHEET DNA BINDING PROTEINS

The methionine repressor controls its own gene as well as structural genes for enzymes involved in the synthesis of methionine. It is a protein with 104 amino acids and forms stable dimers in solution. The structure had been determined by Phillips and colleagues, and it consists of two highly intertwined monomers that form a two-stranded antiparallel β sheet with one strand coming from each monomer (Rafferty *et al.*, 1989). This β sheet forms a protrusion on the surface of the molecule. A similar structure has been deduced for the Arc repressor based on NMR studies (Kaptain *et al.*, 1985). Phillips has also solved the structure of the Met repressor bound to a synthetic DNA fragment containing 18 base pairs (S. Phillips, personal communication). The structure of the Met dimer is not changed greatly by binding to the DNA, which is largely in the B conformation. The two-stranded β sheet of the repressor is found in the major groove of the DNA with side chains from the β strands interacting with base pairs within the operator sequences. These interactions are the base sequence-specific interactions. The DNA itself is somewhat kinked in the center of the operator sequence. That has the effect of narrowing the major groove slightly so that it can form closer bonding to the side chains of the β sheet.

The two-stranded β sheet is thus another DNA binding motif which, unlike the others mentioned above, does not use an α helix for recognition but rather an extended polypeptide chain.

The listing of protein structural motifs that are involved in recognizing DNA sequences (Table I) is necessarily incomplete. A number of DNA binding proteins are known for which no structural data is available. One of these is a protein dimer held together by two α-helical segments that contain leucine residues that facilitate the binding. This so-called leucine zipper is also a fairly common recognition motif, but the nature of its interactions with the DNA is yet unresolved.

What we see in the protein–DNA structures is a wide variety of protein structural motifs that are used in interacting with DNA. α helices are used in at least three different orientations in the helix-turn-helix, zinc finger, or Eco RI recognition motifs. Likewise, the β sheet is used as well. In surveying these structures, one is struck by the variety of recognition motifs that are used. It is likely that more will be discovered in the future, although the total number may not become extremely large since there are probably only a finite and fairly small number of ways in which protein conformations can interact stably with DNA in the double-helical form.

PROTEIN RECOGNITION OF A-TYPE DOUBLE HELICES: tRNA–PROTEIN INTERACTIONS

As pointed out above, the double-helical A conformation has a deep narrow major groove and a flat wide minor groove. The entrance to the major groove is too small for amino acid side chains of protein to enter. The phosphate groups on opposite strands are approximately 4 Å apart, and this suggests that protein recognition of an A-type helix will not take place in the major groove. A-DNA probably occurs infrequently in biological systems. However, the normal conformation of double-helical RNA is an A conformation, virtually identical to that seen in double-stranded A-DNA (Fig. 1, B).

Sequence-specific recognition of RNA molecules by proteins is an integral part of biological systems. How does that occur? We have the most information at present in the system of transfer RNA recognition. Transfer RNA plays a central role in protein synthesis. Transfer RNA molecules have over 75 nucleotides, and they have an anticodon triplet of bases that interacts with the messenger RNA. It forms a stable three-dimensional structure that we know a great deal about through x-ray diffraction analysis of crystalline tRNA molecules (Rich and RajBhandary, 1976). All transfer RNA molecules can be organized in the form of a cloverleaf diagram that contains four stem regions and three loop regions, as shown in Fig. 8. The three-dimensional form of the tRNA molecule is also shown diagrammatically in Fig. 8, where the cloverleaf takes the form of an L-shaped molecule with the three anticodon bases at one end of the L and the amino acid acceptor end at the other end of the L. In protein synthesis, tRNA molecules bind to

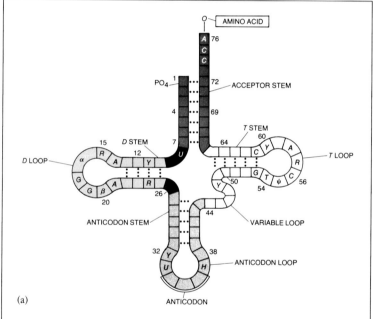

(a)

FIGURE 8a. A cloverleaf diagram of the two-dimensional folding pattern of transfer RNA molecules. This was first deduced in 1965, and this diagram fits all of the several hundred tRNAs that have been sequenced since then. Nucleotide bases found in the same position of all tRNA sequences are indicated, and the ladderlike stems are made up of complementary bases. Abbreviations: R (adenosine or guanosine), Y (cytidine or uridine), T (ribothymidine), ψ (pseudouridine), H (modified adenosine or guanosine).

messenger RNA through the anticodon triplet, and the amino acids are sequentially polymerized into proteins.

The cloverleaf diagram is reorganized considerably in its three-dimensional manifestation. The acceptor stem and the T stem form a continuous RNA double helix, while the D stem and the anticodon stem are stacked upon each other to form another double helix. The molecule itself is largely constructed out of two RNA double helices oriented more or less at right angles to each other. A series of complex interactions among its bases, many tertiary in nature, are used to maintain the molecule in this conformation.

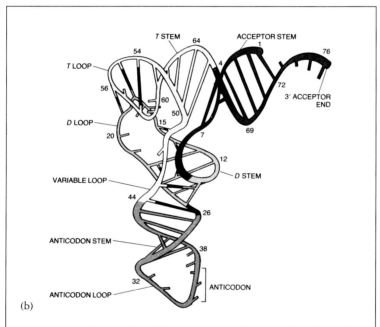

(b)

FIGURE 8b. Shows the folding pattern of yeast phenylalanine transfer RNA as deduced from x-ray diffraction studies. The sugar phosphate backbone of the molecule is represented as a tube, and the crossrungs stand for nucleotide base pairs in the stem region. The short rungs indicate bases that are not involved in base–base hydrogen bonding. (From A. Rich and S. H. Kim (1978), *Scientific American* **238**, 52–62, with permission.)

One of the central events in molecular biology is the aminoacylation of transfer RNA molecules. Each amino acid has its own transfer RNA or group of transfer RNA molecules to which it can be attached. The attachment process takes place on a tRNA aminoacyl synthetase, a large enzyme, one of which is used specifically for each amino acid. The fidelity of transmission of genetic information from nucleic acids to proteins depends upon two sets of events. One is the synthesis of messenger RNA from DNA, in which the fidelity of information transfer is related to the exact placement of messenger RNA bases copied from one strand of a DNA template. The second event is the exact positioning of transfer RNA molecules on the messenger RNA, where the anticodon bases specifically recognize three adjacent nucleotides on the message, and in this way position amino acids in a specific sequence that is necessary for protein synthesis. Aminoacylation of transfer RNA molecules is thus a highly specific step and requires a reliable protein–nucleic acid recognition system.

This recognition problem is made more acute by the observation that the major groove of the RNA double helix is largely inaccessible to the amino acid side chains of proteins, thus precluding the kind of recognition we have seen earlier that occurs between proteins and DNA. The exposed part of the double helix is largely the minor groove, which contains very little information, save for an amino group in the 2 position of guanine that is found on the dyad axis of the double helix. This suggests that other types of recognition systems are operative. In particular, one might anticipate that protein–RNA recognition will largely occur in single-stranded regions of RNA where the individual bases are accessible, as well as in double-helical segments in which there is a modification of the usual Watson–Crick base pairing. In RNA structures, G·U base pairing is found to occur occasionally, and this would modify the conformation of the double helix. It could be a unique recognition element that would be detected in the minor groove of the RNA double helix. Another type of recognition that might also be found is interactions at the very ends of a double-helical segment, since the sequence-specific elements in the major groove would be accessible at the end of a double helix, even though they are inaccessible in the middle.

When we look at the three-dimensional structure of transfer RNA molecules, we can see that most of it is composed of double helical segments, and this would lead us to wonder how recognition could be carried out by the aminoacyl tRNA synthetases. Shortly after the yeast phenylalanine transfer RNA structure was solved, a suggestion was made that the interaction with the tRNA amino acyl synthetases is likely to occur along the inside of the L-shaped molecule shown in Fig. 8 (Rich and Schimmel, 1977). This suggestion has been largely borne out by the structure of two aminoacyl tRNA synthetase–tRNA complexes that have been solved by x-ray diffraction analysis.

The structure of the *E. coli* glutamine tRNA synthetase was solved complexed with tRNA^{Gln} and ATP at 2.8 Å resolution by Steitz and colleagues (Rould *et al.*, 1989). More recently, the structure of yeast aspartic tRNA synthetase complexed to tRNA^{Asp}[7] was solved by Moras and colleagues (Ruff *et al.*, 1991). In both of these, the tRNA molecule has undergone a small conformational change compared to the conformation of yeast phenylalanine tRNA (Kim *et al.*, 1974). Figure 9a shows a comparison of the backbone of yeast tRNA^{Phe} and *E. coli* tRNA^{Gln} superimposed upon each other. It can be seen that both molecules have a fairly similar contour, save for a modification at the acceptor end of the tRNA^{Gln} when it is complexed to the enzyme, as well as a slight change in the anticodon region. Likewise, the backbone of yeast tRNA^{Asp} and yeast tRNA^{Phe} are very similar to each other (Fig. 9b) except for a change in the position of the anticodon loop of tRNA^{Asp}, which is tilted over slightly relative to yeast tRNA^{Phe}.

A number of genetic studies have been done to determine the so-called identity elements in the tRNA that are responsible for its uniqueness. These studies on the specificity of charging of tRNAs have shown that a relatively small number of nucleotides are important for the ability of synthetases to select their cognate tRNAs. In the case of the Gln synthetase, the major identity determinants that have been discovered from these genetic studies show that U35 in the anticodon loop (Schulman and Pelka, 1985), as well as G73 and the base pair 1–72 in the acceptor stem, seem to be important for recognition (Hooper *et al.*, 1972). The crystal structure of the tRNA^{Gln}–synthetase

---

[7] Asp = aspartic acid.

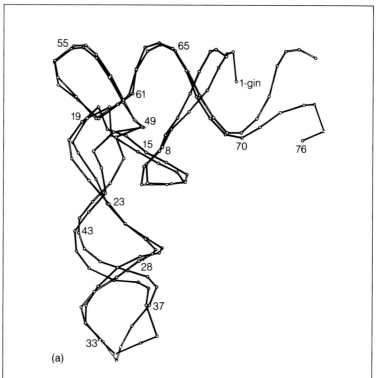

FIGURE 9a. The backbone of yeast tRNA^Asp and *E. coli* tRNA^Gln is shown in the conformations that they have complexed to their respective synthetases in comparison to the backbone conformation of yeast tRNA^Phe crystallized by itself. The diagram on the left compares yeast tRNA^Phe (open bars) with tRNA^Gln (closed bars). The numbers refer to the position of various phosphate groups in the sugar–phosphate backbone. It can be seen that the CCA end of tRNA^Gln is bent down and folded back upon itself. (From M. A. Rould, J. J. Perona, D. Söll, and T. A. Steitz, *Science* **246**, 1135–1142, Copyright © 1989 by the AAAS.)

complex provides strong evidence for the nature of these identity elements and suggests others as well. Figure 10 is a diagram showing a mode of interaction of the tRNA^Gln with its synthetase. It can be seen that the enzyme largely interacts with the tRNA on the inside of the L region. A more detailed view of the acceptor end of the tRNA mole-

(b)

FIGURE 9b. Shows the superposition of tRNA[Asp] (heavy line) and yeast tRNA[Phe] (thin line). It can be seen that the folding of the sugar–phosphate chains is very similar except in the anticodon region at the bottom, where the tRNA[Asp] is moved somewhat to the right. (From M. Ruff, S. Krishnaswamy, M. Boeglin, A. Poterszman, A. Mitschler, A. Podjarny, B. Rees, J. C. Thierry, and D. Moras, *Science* **252,** 1682–1689, Copyright © 1991 by the AAAS.)

cule in the complex is shown in Fig. 11. Here it can be seen that the first base pair, A72–U1, has been disrupted by a side chain from leucine 136 that has wedged between the base pair. Furthermore, the acceptor end of the molecule forms a hairpin with a number of interactions that stabilize it. In particular, this includes interactions with

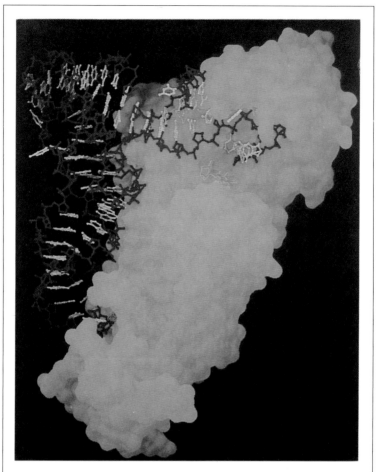

FIGURE 10. Solvent accessible surface representation of the glutamyl aminoacyl synthetase complexed with tRNA^{Gln} and ATP. The tRNA^{Gln} is shown in skeletal representation. The region of contact between tRNA and protein extends across one side of the entire enzyme surface. The acceptor end of the tRNA and the ATP are seen in the bottom of a deep cleft. Protein is inserted between the 5′ and 3′ ends of the tRNA and disrupts the expected base pair between U1 and A72. (From M. A. Rould, J. J. Perona, D. Söll, and T. A. Steitz, *Science* **246,** 1135–1142, Copyright © 1989 by the AAAS.)

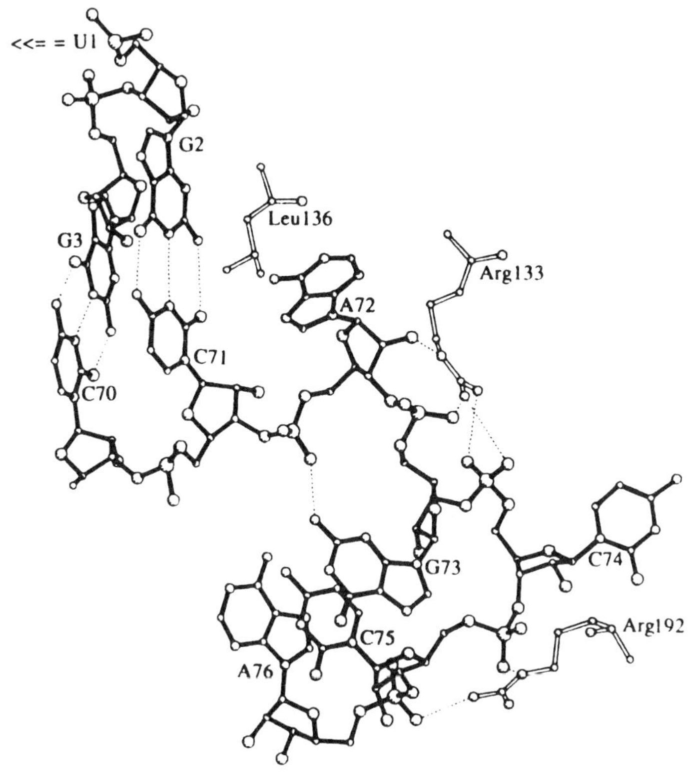

FIGURE 11. The acceptor strand of the tRNA^{Gln} as seen in the synthetase complex. The side chain of leucine (leu) 136 extends from a β turn and wedges between the bases of nucleotides A72 and G2, disrupting the last base pair of the acceptor stem, U1–A72. The enzyme stabilizes the hairpin conformation via the interaction of several basic side chains with the sugar–phosphate backbone. An intramolecular hydrogen bond between the 2-amino group of G73 and the phosphate group of A72 further stabilizes this conformation. (From M. A. Rould, J. J. Perona, D. Söll, and T. A. Steitz, *Science* **246,** 1135–1142, Copyright © 1989 by the AAAS.)

several basic side chains and the sugar phosphate backbone, plus an intramolecular hydrogen bond between the two amino group of G73 and the phosphate group of A72. Only guanine in the 72 position would stabilize this hairpin; thus, it acts as a recognition specificity element not by interacting with the protein, but rather by forming an intramolecular hydrogen bond with another portion of the tRNA.

This also provides a structural basis for the observation that AU or mismatched base pairs are more favored by the Gln synthetase at position 1–72 than GC base pairs (Seong *et al.*, 1989). It is easier to break a mismatched or AU base pair than a CG base pair. There is also a suggestion that two base pairs in the acceptor stem are directly recognized by interactions with the protein and a buried water molecule through the minor groove of the tRNA. Whether this interaction is specific in nature will have to be determined by genetic studies in which these base pairs are inverted, that is, changing a GC for a CG base pair.

Another clear recognition element in this structure is an interaction of uracil 35 in the anticodon region with amino acid side chains involving Glu, Gln and one arginine (Arg) residue (Fig. 12). The interaction of uracil 35 is quite specific, and no other base could be substituted there and still make these interactions (Steitz, 1990). The three anticodon bases are unstacked and splayed out, each binding to a separate region of the protein. The hydrogen bonding to guanine 36 also appears to be specific for that base, and arginine seems to be involved in the interaction with guanine 36 in specifying its identity.

Another example of tRNA recognition by synthetase is seen in the structure of the complex of yeast aspartic tRNA synthetase complexed to its cognate tRNA (Ruff *et al.*, 1991). This structure involves only a slight distortion of the tRNA conformation when it is bound to the synthetase compared to its unbound state. In the unbound state, the tRNA^{Asp} conformation is very similar to that of tRNA^{Phe}, and as shown in Fig. 9b, the anticodon region of the tRNA^{Asp} is in a slightly different position, rotated towards the protein. In the case of yeast tRNA^{Asp}, genetic studies have been done to locate the identity units that are specific for its function. In particular, the positions of the major determinants for aspartylation by the yeast synthetase are shown in the

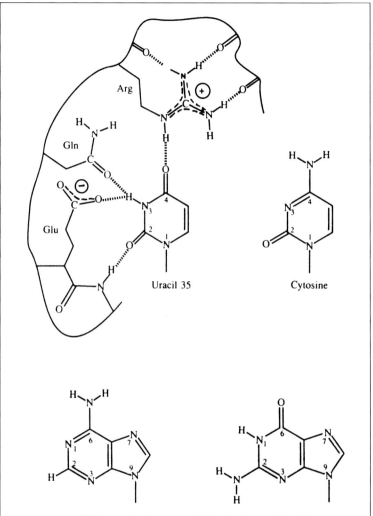

FIGURE 12. Schematic drawing of the hydrogen-bonding inter-actions between U 35 of tRNA^{Gln} and the Gln aminoacyl synthe-tase. The protein-binding pocket specifies uracil rather than cytosine by virtue of the ionic hydrogen bonds. Adenine (lower left) and guanine (lower right) would not fit into the binding site. (From T. A. Steitz (1990), *Quart. Rev. Biophys.* **23**, 205–280, with permission from Cambridge University Press.)

shaded brackets of Fig. 13 (Pütz *et al.*, 1991). These consist of a G residue, the so-called discriminator base immediately adjacent to the acceptor stem, a GU base pair at the end of the D stem, and the three anticodon bases. All of these regions are in close contact with the protein, as seen in the crystal structure analysis. Furthermore, it has been shown that the sequence of yeast tRNA^Phe can be used as a basis for changing it so that it will charge with aspartic acid. This is carried out by making the four changes indicated in Fig. 13. Thus, two of the anticodon bases have been changed so that they now code for the aspartic

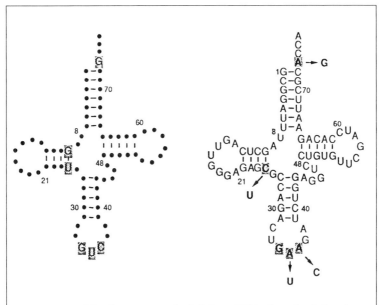

FIGURE 13. The diagram on the left is a tRNA cloverleaf showing the position of the major determinants for aspartylation by the yeast tRNA^Asp aminoacyl synthetase as deduced by mutational analysis. The sequence on the right is that of yeast tRNA^Phe in which the Asp determinants, indicated by arrows, have been integrated into the shaded position. These four changes were enough to make a molecule that could be actively aminoacylated with Asp. (From J. Pütz, J. D. Puglisi, C. Florentz, and R. Giegé, *Science* **252,** 1696–1699, Copyright © 1991 by the AAAS.)

acid tRNA, and finally the GC base pair at the end of the D stem has been changed to a GU base pair. As mentioned above, the GU base pair is held together by a wobble pairing that has the effect of thrusting the 2-amino group of G further out into the minor groove of the RNA double helix, where it is readily accessible to amino acid side chains. Substitution of a GC base pair for a GU base pair thus produces a significant change in the conformation of the RNA double helix, and as such, it becomes a recognition element by the protein. In the structure of the complex, the anticodon loop undergoes an important conformational change that results in unstacking the bases of the anticodon so that it can bind to the protein. The details of the discriminating interactions are not available as yet, but they may involve interactions similar to those seen in the yeast tRNA[Gln] complex.

SYNTHETASE RECOGNITION INVOLVING ACCEPTOR STEMS ONLY

In many tRNAs such as the two whose structure has been solved complexed to their synthetase, protein interaction with the anticodon appears to be essential in recognizing the tRNA. However, this is not true of tRNAs in general. For some time it has been known that tRNAs are found in which the anticodon can be removed and there is still efficient charging by an amino acid to the remainder of the molecule. Thus, we are led to ask what is the nature of the discriminating interactions that define the identity of the tRNA. The most striking example of discriminator information is seen in the work of Schimmel and colleagues in the *E. coli* alanyl tRNA system. The tRNA[Ala 8] is unique in that it has at the third base pair position in the acceptor stem a GU base pair rather than a normal Watson–Crick base pair. As mentioned above, the GU base pair provides a considerable uniqueness in its conformation. The remarkable discovery has been made that efficient charging of tRNA[Ala] fragments can be obtained using a minihelix, which consists of only the acceptor stem and the T stem (Schimmel, 1991). In addition, even a microhelix can also be charged containing only the acceptor stem base pairs, as shown in Fig. 14. It has been demonstrated that other tRNAs that do not have a GU base pair in the

[8]Ala = alanine.

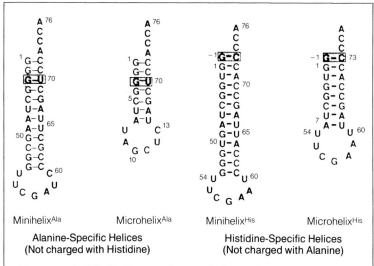

FIGURE 14. Minihelices and microhelices that are aminoacylated specifically with either alanine or histidine by their respective aminoacyl synthetases. A critical base pair for aminoacylation specificity and efficiency is highlighted in each case. (From P. Schimmel (1991) *FASEB J.*, **5**, 2180–2187, Copyright © 1991 from *FASEB J.*)

third position in the anticodon stem can be converted to charging by the alanyl synthetase by simply changing the third base pair in the acceptor stem to a GU base pair. Thus, it seems that the GU not only is necessary, but appears to be sufficient for aminoacylation by the alanyl enzyme. In further work it has been shown that the amino group on the 2 position of guanine is essential for recognition, since if inosine lacking a 2-amino group is substituted in this position, the enzyme will no longer charge the minihelix or the microhelix.

Another example of uniqueness of a single determinant is found in the charging of histidine tRNA. The tRNA[His] has a base pair at the -1 position, that is, a base pair involving the discriminator base with an additional nucleotide on the 5′ end. Minihelices and microhelices containing this additional GC base pair are charged specifically by the histidine synthetase (Schimmel, 1991). The charging sequences are shown in Fig. 14. Furthermore, tRNA[Ala] has been modified so that it has an additional GC base pair in this position, and this is enough to convert the tRNA so that it can be charged by the histidine enzyme. This is another example of a recognition element that is both necessary and sufficient for recognition by the histidine synthetase. Although structural studies have not been done, it should not be difficult for proteins to discriminate between GC base pairs at the ends of double helical segments, since those are accessible in the major groove side at the end of the helix. However, the nature of this recognition will have to await structural analysis.

There is much less structural data available in the field of protein–RNA recognition. However, a number of genetic and footprinting studies have been carried out in other systems that support the general thesis that recognition of RNA molecules is generally found at regions that have single-stranded character or unusual conformations. A good example of this is seen in the HIV protein, TAT, that binds to an RNA segment, the transactivating response element (TAR). TAR is a very large RNA stem loop structure that has a bulge in it near the loop end. The major recognition elements for the TAR protein are found in the bulge region (Weeks *et al.*, 1990), again reinforcing the concept that single-stranded regions or segments in which the RNA double helix is disrupted represent recognition opportunities for proteins.

Z-DNA RECOGNITION

Z-DNA is a higher-energy conformation of the B-DNA double helix in which there is a considerable reorganization of the double helix so that it is now in the left-handed mode. A-DNA is also a somewhat higher-energy form, a less stable conformation of the B-DNA double helix. However, the energy required to go from B to A is much less than that required to go from B to Z. The energy of Z-DNA is much higher than that of B-DNA, and normally it is stabilized in biological systems by negative supercoiling (Peck *et al.*, 1982) or proteins that bind to it.

In some respects, the Z-DNA conformation is just the opposite of the A-DNA conformation as far as recognition is concerned. In A-DNA, the major groove is deep, and access to it is blocked by the fact that the two sugar phosphate backbones are close together, with about 4 Å separating the phosphate groups. This precludes amino acid side chains recognizing the major groove, while the minor groove is accessible on the outside. However, in Z-DNA what corresponds to the minor groove is very narrow and deep and only has room in it for water molecules. The major groove is now converted to the convex outer surface of the molecule, so that all of the specificity elements of DNA base pairs are found on the outer surface of Z-DNA where they are fully accessible.

Every other residue along the Z-DNA chain has the base in the *syn* conformation, that is, rotated relative to the sugar (Wang *et al.*, 1979). In A-DNA and B-DNA, all the bases are in the *anti* conformation. Since purines rotate to the *syn* position more readily than do pyrimidines, it means that Z-DNA–forming sequences form more readily where there are alternations of purines and pyrimidines.

Although Z-DNA binding proteins have been studied, none of them has been purified to the point where they have been crystallized either alone or with segments of Z-DNA. Thus, we do not have information about the structural basis for Z-DNA recognition. However, a great deal is known about Z-DNA and its interaction with a number of small ligands, specifically cations.

The B to Z transition has been studied most frequently with poly (dG–dC), and the influence of various cations on the B to Z transition has been studied extensively. Poly (dG–dC) converts to Z-DNA in 5M NaCl. Mg^{2+} is an order of magnitude more effective in stabilizing the Z conformation than is the sodium ion (Behe and Felsenfeld, 1981). However, the cation cobalt hexaamine $Co(NH_3)_6^{3+}$ is five orders of magnitude more effective than the sodium ion. Thus, very small concentrations of cobalt hexaamine will stabilize poly (dG–dC) in the Z conformation. The structural basis of this stabilization was revealed in an x-ray diffraction analysis of $d(CG)_3$ carried out at 1.5 Å resolution (Gessner *et al.*, 1985). The results of this analysis are shown in Fig. 15, in which the positions of the magnesium hexahydrate and cobalt hexaamine complexes are shown. Figure 15 also shows in greater detail the interaction of the cobalt hexaamine ion with Z-DNA. It can be seen that three of the amino groups of the cobalt hexaamine ion make specific hydrogen bonds, both to O6 and N7 of a guanine residue and to a neighboring phosphate oxygen. It is this highly specific assembly of five hydrogen bonds that strongly stabilizes DNA in the Z conformation. It does so not only by stabilizing the Z conformation through hydrogen bonding, but also through the electrostatic stabilization of a

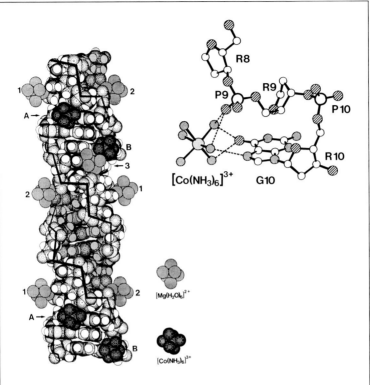

FIGURE 15. *Left:* Z-DNA with its attached ions in the magnesium/cobalt hexaamine crystal. Two types of octahedral complexes are shown on the surface of the molecule. *Right:* Coordination of cobalt hexaamine to Z-DNA. The ammonia molecules of the cobalt hexaamine complex are found hydrogen-bonded to the phosphate group P9 and to N7 and O6 of guanine 10. This shows the coordination of complex A at the left. (From R. V. Gessner, G. J. Quigley, A. H.-J. Wang, G. A. van der Marel, J. H. van Boom, and A. Rich, *Biochemistry* **24**, 237–240, Copyright © 1985 American Chemical Society.)

closely appended ion with three positive charges that is linked to the molecule.

There is no information about the interaction of amino acid side chains with Z-DNA, although it has been reported that polyarginine is very effective in stabilizing poly (dG–dC) in the Z conformation

(Klevan and Schumaker, 1982). In view of the structural information seen in the cobalt hexaamine complex, it would not be surprising to find that positively charged side chains such as arginine are influential in stabilizing the Z conformation when the structure of a Z-DNA binding protein attached to Z-DNA has been determined.

Conclusions

This overview started with an analysis of the sequence-specific informational constraints found in the nucleic acid double helix. In particular, it stressed the fact that the sequence-specific residues are found largely (if not entirely) in the major-groove side of the Watson–Crick base pair. The minor groove offers a means of differentiating CG from AT (or AU) base pairs, but does not seem to be capable of having specific interactions. In the course of this analysis, it was pointed out that recognition by amino acid side chains forming two hydrogen bonds to bases would represent an effective way of recognizing specific sequences.

A number of studies have been carried out with protein–DNA recognition that clearly point out that the major groove is the important part of the DNA double helix that is involved in recognition. However, the number and types of recognition interactions are extremely diverse, and there is no simple system that can be used to describe protein recognition of DNA. There is perhaps a slight exception to this seen in the interaction of the zinc fingers with DNA. In that complex, five of the six base pairs are recognized by specific interaction between arginine and guanine.

In general, a diverse number of interactions are found in sequence-specific recognition, many involving more than one hydrogen bond to bases, but some involving just single hydrogen bonds to bases. It is not clear that the latter are specific in their recognition. Frequently thymine is recognized by the hydrophobicity of its methyl group. In addition, there are a large number of interactions between the DNA backbone and proteins that are found in these complexes. Changes in nucleic acid conformation appear to be a part of some recognition systems.

The situation regarding the A-type double helix as found in RNA is in some ways more simple and more diverse. It is simpler because the major groove is largely blocked, inaccessible to side chains; therefore, most recognition is found in single-strand regions or in regions of the double helix in which modifications of the Watson–Crick base pairing, as in the GU base pair, produce some unique feature that can be used as a recognition device. A number of different interactions are seen with single-stranded regions that are largely responsible for recognition.

There is thus considerable diversity in the types of interactions between proteins and nucleic acids, involving a large number of protein motifs. In a way, this system works in a manner that is probably in concordance with the expectations of Linus Pauling at the beginning of this field some 45 years ago.

ACKNOWLEDGMENT

This research was supported by grants from the National Institutes of Health, the National Science Foundation, and the Office of Naval Research.

REFERENCES

Aggarwal, A. K., Rodgers, D. W., Drottar, M., Ptashne, M., and Harrison, S. C. (1988). *Science* **242,** 99–107.

Anderson, J. E., Ptashne, M., and Harrison, S. C. (1987). *Nature* **326,** 846–852.

Anderson, W. F., Ohlendorf, D. H., Takeda, Y., and Matthews, B. W. (1981). *Nature* **290,** 754–758.

Behe, M., and Felsenfeld, G. (1981). *Proc. Natl. Acad. Sci. USA* **78,** 1619–1623.

Brennan, R. G., Roderick, S. L., Yoshinori, T., and Matthews, B. W. (1990). *Proc. Natl. Acad. Sci. USA* **87,** 8165–8169.

Davies, D. R. (1967). *Annu. Rev. Biochem.* **36**, 321–364.

Day, R. O., Seeman, N. C., Rosenberg, J. M., and Rich, A. (1973). *Proc. Natl. Acad. Sci. USA* **70**, 849–853.

Felsenfeld, G., Davies, D. R., and Rich, A. (1957). *J. Am. Chem. Soc.* **79**, 2023–2024.

Felsenfeld, G., and Rich, A. (1957). *Biochim. Biophys. Acta* **26**, 457–468.

Gessner, R. V., Quigley, G. J., Wang, A. H.-J., van der Marel, G. A., van Boom, J. H., and Rich, A. (1985). *Biochemistry* **24**, 237–240.

Hooper, M. L., Russell, R. L., and Smith, J. D. (1972). *FEBS Lett.* **22**, 149.

Jordan, R. S., and Pabo, C. O. (1988). *Science* **242**, 893–899.

Kaptain, R., Zuiderweg, E. R. F., Scheek, R. M., Boelens, R., and van Gunsteren, W. F., *J. Mol. Biol.* **182**, 179–182.

Kim, S. H., Suddath, F. L., Quigley, G. J., McPherson, A., Sussman, J. L., Wang, A. H.-J., Seeman, N. C., and Rich, A. (1974). *Science* **185**, 435–440.

Kim, Y., Grable, J. C., Love, R., Greene, P. J., and Rosenberg, J. M. (1986). *Science* **234**, 1526–1541.

Kissinger, C. R., Liu, B., Martin-Bianco, E., Kornberg, T. B., and Pabo, C. O. (1990). *Cell* **63**, 579–590.

Klevan, L., and Schumaker, V. N. (1982). *Nucl. Acids Res.* **10**, 6809–6817.

Lawson, C. L., Zhang, R.-G., Schevitz, R. W., Otwinowski, Z., Joach-imiak, A., and Sigler, P. B. (1988). *Proteins* **3**, 18–31.

Lee, M. S., Gippert, G. P., Soman, K. V., Case, D. A., and Wright, P. E. (1989). *Science* **245**, 635.

Luisi, B. F., Xu, W. X., Otwinowski, Z., Freedman, L. P., Yamamoto, K. R., and Sigler, P.B. (1991). *Nature* **252**, 497–505.

McClarin, J. A., Frederick, C. A., Wang, B.-C., Greene, P., Boyer, H. W., Grable, J., and Rosenberg, J. M. (1986). *Science* **234,** 1526–1541.

McKay, D. B., and Steitz, T. A. (1981). *Nature* **290,** 744–749.

O'Brien, E. J. (1967). *Acta Crystallogr.* **23,** 92–106.

Otwinowski, Z., Schevitz, R. W., Zhang, R.-G., Lawson, C. L., Joachimiak, A., Marmostein, R. Q., Luisi, B. F., and Sigler, P. B. (1988). *Nature* **335,** 321–329.

Pauling, L. (1946). *Chem. and Eng. News* **24,** 1375–1377.

Pavletich, N. P., and Pabo, C. O. (1991). *Science* **252,** 809–816.

Peck, L. J., Nordheim, A., Rich, A., and Wang, J. C. (1982). *Proc. Natl. Acad. Sci.* **79,** 4560–4564.

Pütz, J., Puglisi, J. D., Florentz, C., and Giegé, R. (1991). *Science* **252,** 1696–1699.

Rafferty, J. B., Somers, W. S., St.-Girons, I., and Phillips, S. E. V. (1989). *Nature* **341,** 705–710.

Rich, A., and Kim, S. H. (1978). *Sci. Amer.* **238,** 52–62.

Rich, A., and RajBhandary, U. L. (1976). *Ann. Rev. Biochem.* **45,** 805–860.

Rich, A., and Schimmel, P. R. (1977). *Nucl. Acids Res.* **4,** 1649.

Robertus, J. D., Ladner, J. E., Finch, F. T., Rhodes, D., Brown, R. S., Clark, B. F. C., and Klug, A. (1974). *Nature* **250,** 546–551.

Rosenberg, J. M., Seeman, N. C., Kim, J. J. P., Suddath, F. L., Nicholas, H. B., and Rich, A. (1973). *Nature* **243,** 150–154.

Rosenberg, J. M., Seeman, N. C., Day, R. O., and Rich, A. (1976). *J. Mol. Biol.* **104,** 145–167.

Rould, M. A., Perona, J. J., Söll, D., and Steitz, T. A. (1989). *Science* **246,** 1135–1142.

Ruff, M., Krishnaswamy, S., Boeglin, M., Poterszman, A., Mitschler, A., Podjarny, A., Rees, B., Thierry, J. C., and Moras, D. (1991). *Science* **252,** 1682–1689.

Sauer, R. T., Jordan, S. R., and Pabo, C. O. (1990). *Advances in Protein Chem.* **40,** 1–61.

Schevitz, R. W., Otwinowski, Z., Joachimiak, A., Lawson, C. L., and Sigler, P. B. (1985). *Nature* **317,** 782–786.

Schimmel, P. (1991). *FASEB J.* **5,** 2180–2187.

Schulman, L. H., and Pelka, H. (1985). *Biochemistry* **24,** 7309–7314.

Schultz, S. C., Shields, G. C., and Steitz, T. A. (1990). *J. Molec. Biol.* **213,** 159–166.

Seeman, N. C., Rosenberg, J. M., Suddath, F. L., Kim, J. J. P., and Rich, A. (1976a). *J. Mol. Biol.* **104,** 109–144.

Seeman, N. C., Rosenberg, J. M., and Rich, A. (1976b). *Proc. Natl. Acad. Sci. USA* **73,** 804–808.

Seong, B. L., Lee, C.-P., and RajBhandary, U. L. (1989). *J. Biol. Chem.* **246,** 6504.

Steitz, T. A. (1990). *Quart. Rev. Biophys.* **23,** 205–280.

Wang, A. H.-J., Quigley, G. J., Kolpak, F. J., Crawford, J. L., van Boom, J. H., van der Marel, G., and Rich, A. (1979). *Nature* **282,** 680–686.

Watson, J. D., and Crick, F. H. C. (1953). *Nature* **171,** 737.

Weeks, K. M., Ampe, C., Schultz, S. C., Steitz, T. A., and Crothers, D. M. (1990). *Science* **249,** 1281–1285.

Wharton, R. P., and Ptashne, M. (1985). *Nature* **316,** 601–605.

Wolberger, C., Dong, Y., Ptashne, M., and Harrison, S. C. (1988). *Nature* **335,** 789–795.

4

THE IMPACT OF LINUS PAULING ON MOLECULAR BIOLOGY: A REMINISCENCE

———◆———

Francis Crick
The Salk Institute

With great pleasure, I would like to reminisce about Linus Pauling's impact on molecular biology. I shall not write about his impact on chemistry as a whole, as that has been described elsewhere by a number of authors in this book.

My copy of *The Nature of the Chemical Bond* was not the first edition, but the second 1945 one, in fact the 1946 printing. I gave Jim Watson a copy in 1951 as a present for Christmas. I also bought for myself the first (1949) edition of *General Chemistry*. I had done no chemistry as an undergraduate, although I had done a little in high school, and it was, as Max Perutz described in Chapter 2, a mixture of things that you had to learn by heart. So it was an eye-opener for me to read Linus Pauling's book. I must confess: I didn't read all the parts of inorganic chemistry but I devoured most of the organic chemistry. It is almost true to say that's the only chemistry I ever learned.

Now we come to the proper subject of this chapter: You don't really have to be told that chemical bonds, loosely speaking, can be divided into two sorts: the strong bonds (the homopolar bonds and the ionic bonds) and the weak bonds (the hydrogen bonds and the van der Waals forces). As I understand it, chemists at the time were really more interested in the strong bonds. What Linus Pauling did was to point out that for many purposes, and especially for biological molecules, it was the weak bonds that we had to pay attention to, partly because there are usually many of them and their effects add up.

Linus Pauling's interests were not only in chemistry but also in what he called physiology. Today we would probably call it molecular biology. He therefore initiated a program for finding the exact bond distances and angles in certain small organic molecules to an accuracy of about 0.01 Å. For example, for a single carbon carbon bond the distance is about 1.54 Å. This distance depends on the nature of the bond, whether it is single, double, or has some double-bond character. These distances and angles could be studied using x-ray diffraction. As you already know, in those days, it wasn't an easy matter to solve structures of crystals by x-ray diffraction in the way they do nowadays. It wasn't something you could embark on easily and involved quite a lot of work even to solve one small structure.

What was needed were the general rules for the character of the hydrogen bond: that the NH–O bond was longer than the OH–O bond and the distances and various angles of approach, and so on. Also, something needed to be found about the van der Waals distances, which of course make what you might call a "snug fit" but are not terribly directional. These estimates were all guided by a knowledge of quantum mechanics, but usually in a more rule-of-thumb form rather than by exact calculations.

While I was still in the Admiralty—I don't know exactly when it was, but I left the Admiralty in 1947—I came across an article in a journal that passed over my desk concerning a paper in *Chemical and Engineering News*. I've recently seen this paper and I don't think I then read the original but a reference to it. It stressed the importance of the hydrogen bond. At that time I didn't know what a hydrogen bond was. And it was written by somebody I had never heard of (being a physicist in those days) called Linus Pauling. This seemed rather an unusual name and I determined to know more about him. I've recently had the opportunity to read that article because Alex Rich kindly sent me a copy, or rather he sent me a published reprint of this article. The original was in *Chemical and Engineering News* in May, 1946 (**24**(10), 1375–1377). This reprint was published in the *Journal of NIH Research* (July, 1990; **2**, 59–64) as a "landmark paper." It also has a present-day commentary by Pauling, which I found very good reading, and also a commentary making a number of interesting points by Alex Rich himself.

You must realize that the original article was published in 1946, before Pauling had even started on the alpha helix, long before the DNA structure, and very long before many of the chapter topics included in this book. He pointed out the importance of the size, and especially the shape, of molecules and in particular organic molecules. It was already known that there was a difference in the behavior of L and D versions of the same molecule. In other words, if you had a small organic molecule and you wanted to see how it would react in a biological situation, and you compared one optical isomer of that molecule with its mirror image, you usually found that they didn't produce the same physiological results. But I don't think people at that time were thinking about the exact size and shape of biological molecules. The other point that he stressed was that that interaction was likely to be complementary rather than like-with-like. Of course that idea—the idea of lock and key—goes back to the beginning of the century, but not everyone believed it and it was important to reemphasize it. There is a paper by Pauling and Delbrück, published in 1940 opposing the idea of another theorist—Pascual Jordan—who said there could be a like-with-like attraction. These matters are very important if you are interested in things like biological replication. Pauling and Delbrück stressed that it was more likely to be complementary.

Jim Watson and I were aware of that article and we did in fact quote it in one of our papers in 1953. There was an even more remarkable suggestion by Linus, which he made some time in the late 1940s—that the gene might consist of two mutually complementary strands. Jim and I didn't read that article until after we had discovered the double helix, but there was no doubt that Linus put forward that idea before we did.

The *Chemical and Engineering News* article also pointed to the range of sizes that it was important to know more about. They were in the region between 10 and 100 Å, and he predicted that this range would be important for biology and that it was a region that was largely unexplored at that time.

Let me now go through a number of the more detailed ideas. Very early (in 1936) in an article with Mirsky, Pauling had considered the general problem of protein structure. They suggested the essential nature of protein denaturation: that in protein denaturation, generally speaking, the strong bonds are unaffected and the weak bonds (because of the thermal jostling that you get when you denature a protein, say by heat) are disturbed so that the whole structure is jumbled. Because the integrity of the structure depends on a large number of weak bonds you tend to get a rather rapid change—it's a cooperative phenomenon—as you raise the temperature. You must remember that in those days there was no sequence data for any protein. Indeed, there were people who didn't believe that a protein had an exact amino acid sequence. Many people thought that proteins were "colloids," a rather ill-defined term. They didn't think that each protein had a characteristic molecular weight. They only had a rough idea of their size and shape and that for only a few of the common proteins. So really rather little was known about protein structure when Pauling and Mirsky made these suggestions which are certainly along the right lines.

There had been of course some pioneer x-ray diffraction work by Astbury, mainly on fibers and on nails, hairs, and things of that sort. He had found two sorts of x-ray diffraction patterns: the alpha-type and the beta-type. He had produced a model for the beta sheet which we now know was crudely right. He produced a model for the alpha turn which was hopelessly wrong. It was because of all this, as I understand it, that Pauling started up his own program, so that by 1946, when he wrote the *Chemical and Engineering News* article, he and his colleagues had solved the structure of two amino acids and two small peptides. Even that small amount of data was invaluable in showing things about the nature of hydrogen bonding and about the nature of the peptide bond. Of course it was much helped by Pauling's deep knowledge of chemistry, so that from a relatively small amount of data he could often derive the correct conclusions and extend those conclusions.

He was very interested in another subject at that time. That was the question of antibodies. Again that's essentially the idea of complementarity, that antibodies would be complementary in shape (I use "shape" in a loose sense here) to their antigens. I think other people had had this idea but with a number of colleagues, especially Dan Campbell, he did a lot of experimental work to show that the idea was certainly on the right lines. The article makes an interesting remark: Complementarity is rather rare in chemistry except for one case, and that is in crystallization. It is true that his interest in antigens led him to a theory of antigen formation which we now know is not correct, but it certainly stimulated people to arrive at the clonal selections theory, which is the right one.

There is another topic that comes up in the article, and it's really very farsighted, although nowadays people have begun to take it for granted. It concerns the question of enzyme action. Pauling says in the article that he realized that enzyme action could go both ways, and therefore the most important thing the enzyme had to do was to lower the activation energy. And therefore that the shape of the site (again I'm using "shape" in a broad sense) would interact with the transition state that happens during the enzyme action. This idea, I believe, was somewhat lost sight of for a number of years but has very much come to the fore in recent times, and people are now designing antibodies that can act as enzymes, using this very idea.

Now let us come to the molecular structures. There was a whole series of them that came out on each other's heels—the pleated sheet was a distinct improvement on Astbury's which was essentially a planar structure. Pauling realized that it would be pleated. Then there was the alpha helix which he tells us he first thought of in Oxford, when he was a visiting professor, and was in bed with a cold, remarking that those were the days before vitamin C. I was by that time in the physics lab in Cambridge, working as a graduate student under Perutz. Kendrew was also there and we were under Bragg's loose supervision. Bragg, Kendrew, and Perutz tried to build models of the polypeptide chain that would fit the Astbury alpha pattern, not knowing that Pauling was also trying. They did it really rather sensibly and carefully, but they made two mistakes. They didn't realize that the peptide bond would be planar. I must say I think that was partly bad luck because they did ask a theoretical chemist and got the wrong advice. The other thing is a rather technical matter. On the meridian of the alpha x-ray pattern there is a reflection at about 5.1 or 5.15 Å. That would normally mean—in fact it really does mean when you get right down to it—that the structure had an integer screw axis. Now we know that in crystallography, leaving aside all those funny structures that Pauling has written about, you only have repeating rotational and screw axes which are two-, three- or sixfold. Bragg realized that you didn't have these restrictions for an isolated helix. He was quite prepared to have, say, a fivefold or for that matter I suppose a sevenfold axis. But they tried to make it an integer helix because of that reflection on the meridian. We now know that that is misleading. The reason is that the alpha helices in those biological materials that give the x-ray pattern are not running side-by-side. Because alpha helices tend to pack together at an angle, they are coiled round each other, as indeed Pauling and I pointed out independently a little later when the alpha helix was understood. So Bragg, Kendrew and Perutz did not discover the alpha helix. All the structures they built looked wrong. In a sense, one of their models was an alpha helix (which has 3.6 residues per turn), but they twisted the poor thing to have four residues per turn, so it looked pretty awful.

What Pauling did was to realize that you didn't have to have an integer screw axis and that the peptide bond was planar, and he produced what was essentially the correct structure. He suggested one or two other structures at the time; for example, there was a gamma helix, but that had too big a hole in the middle. I'll say something in a moment about the general business of determining fiber structures.

I'm sure you wouldn't want me to give a talk without mentioning the structure of DNA. There's now been a movie made about this in which some of us are impersonated by actors. I noticed that Linus Pauling, although he is mentioned very often in the film, only appears offstage. Perhaps they couldn't find any actor that could impersonate him properly! An actor impersonates Peter Pauling who arrives in a rather sporty car—I don't think he actually did arrive in that way.

Many of you will know that Linus Pauling produced a structure for DNA that had three chains, which was incorrect. It was really partly a matter of bad luck because although he was hoping to get some good x-ray patterns himself, he used the old x-ray patterns that Astbury had taken of DNA. We know now that they were a mixture of the two forms: the A and the B form. So he was using data which didn't correspond to any real single structure. In addition, his structure was too dense. He made the same mistake that we had made in our first structure, of not realizing how much water there was there.

There was one feature that I found, I must say, very odd and I don't know what the explanation is. He put a hydrogen atom on the phosphate. This will normally only occur at a pH of about 1. Those hydrogens were forming hydrogen bonds and holding the whole structure together. I was so disturbed by this that I took down my book of organic chemistry to see whether that was the reasonable thing to do. The book was *General Chemistry* by Linus Pauling that I mentioned earlier.

Having said all that, I must say that in addition to giving us a jolt, so that we went on and found the structure—and if we hadn't done it, other people would have done it within quite a short time—apart from that, Pauling had a much more important and profound effect on the solving of the structure of DNA. The reason is the following. It's the nature of the problem you're confronted with. As you know, when you take x-ray data you are getting essentially the three-dimensional Fourier components of the electron density. That's what's doing the scattering. But, as is also well known, you only get half the information. You get the amplitudes of the Fourier components but not their phases—this is the so-called phase problem. Now when you come to build a model you also have to use information—that is to say, you must use bond distances and angles that are plausible. In a popular lecture I always have to explain: That's not enough to define the structure because of course you have rotation about single bonds. Nevertheless there are very many constraints on any possible model. So the x-ray diffraction problem is what mathematicians call an "ill-posed problem" and it can only be solved by constraints of one form or another. The ones used by the computers nowadays are that the electron density shouldn't go negative and, to some extent, that the atoms approximate points. But of course those methods weren't available to us then.

FIGURE 1. At the California Institute of Technology, Francis Crick with Linus Pauling, celebrating Pauling's 85th birthday on February 28, 1986.

Thus, you had two techniques for approaching the problem: one using x-ray diffraction, which only had part of the information you need; and the other using the bond distance and angles, which again had only part of the information you need because you don't know the rotation about single bonds. What Pauling realized was that by putting those two techniques together you could often arrive at the structure, and that you couldn't arrive at it by either of them independently. This is especially true for fibers because the monomers of a polymer have to join on head-to-tail. For a polypeptide chain, since the peptide bond is planar, if you put in all the distances and fixed angles you only have two parameters to define the whole structure, which are the rotations about two of the bonds of the alpha carbon atoms. So it is really a two-parameter problem, and that's why you can show it's correct using a rather limited amount of x-ray data, including that 1.5 Å reflection that Perutz found. Now it was this lesson that we absorbed by studying what Pauling had done to obtain the alpha helix. And it was this lesson that Jim and I tried to impart to Maurice Wilkins and Rosalind Franklin, but Rosalind was reluctant to use that method because she wanted to use the method of Patterson (to which Pauling referred), which is a difficult one. For that reason she concentrated on the A form of DNA which gave much better spots and much more information, and put aside the B form, even though she had produced a picture which essentially gave the game away and showed that the structure was helical. And so it was essentially on the basis of Linus Pauling's ideas that Watson and I were able to solve the structure of DNA.

Linus Pauling made one other contribution which I must mention. That is, when we published the base pairs we weren't sure whether the GC pair had two hydrogen bonds or three. The original paper showed two, but in our Royal Society paper we said it might be two or three. Pauling realized that if you looked at the shape of the G and the C and didn't (as we did, rather naively) accept the experimental data at its face value, but allowed for the fact that there might be errors in it, he was able, using his knowledge of chemistry, to get a slightly better shape for G and C and it was then perfectly clear they could easily form that third bond. We now know of course, from lots of data, that that is correct.

Let me leave DNA and go on to another biological topic to which Linus made a very remarkable and surprising contribution and that concerns the haemoglobin molecule in sickle cell anemia. It was what Linus called a molecular disease. He realized that you only got sickling in the venous blood and not when haemoglobin has oxygen on it, and therefore the disease was almost certainly something to do with haemoglobin. I remember this article with Harvey Itano coming out in 1949. They showed there was a difference in the charge of the haemoglobin molecule, strictly one charge on half the molecule so two charges on the whole molecule. I remember reading this paper (actually while I was trying to centrifuge some haemoglobin) and being extremely struck by it. What was significant was that it had been shown, just about that time, by J. V. Neel, that sickle cell anemia was inherited in a Mendelian fashion. So here we had a case of a mutant Mendelian gene which had made this change in a protein molecule. It wasn't possible in those days to track down exactly what the change was—that was done later in our lab by Vernon Ingram, using Sanger's methods of fingerprinting. I don't think most protein chemists were aware of the implications of Pauling's paper. When it was brought down to the chemical details and they could see the change, this actually altered the whole nature of protein chemistry. Up to that time protein chemists were only vaguely aware (if at all) that their subject had anything to do with genetics. It seems extraordinary to many of you nowadays to believe this, but that was the case, so much so that Fred Sanger asked if we would give lectures on genetics to him and his colleagues. They were given informally in the sitting room of my house in Cambridge. This was the result of the work that was initiated by Pauling on haemoglobin as a molecular disease.

In 1965 he made an even more remarkable contribution, with Zuckerkandl, which we can summarize under the phrase of "the molecular clock." This was thought of at the time as an outrageous idea. They suggested that mutations accumulate with time in evolution, in a particular protein, in a fairly regular manner, and therefore can be used to measure evolutionary time, an idea which we now know to be broadly correct. But at the time it seemed so far-fetched that most people didn't immediately accept it.

With the landmark report there is a commentary by Pauling. I was very struck by one thing he said when he was asked by someone, "How do you get ideas?" He said (and in my experience this is the correct answer), "If you want to have good ideas, you must have a lot of ideas, and throw most of them away." Then he went on to say, "It's a matter somewhat of a sixth sense of intuition which ones you think are the promising ones." Of course that is a step for which one is not always infallible, so that it isn't surprising that some of Pauling's ideas have not turned out to be right. But I believe that's because a man who is always right almost never says anything significant. I think it's a tribute to his fertility that such a high proportion of his ideas were right even if not every one of them was.

Let me try and summarize what I have been saying. The importance of Linus Pauling to molecular biology was that he had the right basic ideas about chemistry. You could say, "Well, couldn't anybody have had those ideas?" It so happens that there was another important figure in molecular biology—Max Delbrück, who was at Cal Tech for many years, who started with a different set of ideas. He was a physicist. The reason he came into biology—the reason he started work on phage—was because he thought that biological replication was such a mysterious process that it would involve new laws of physics. He didn't believe they would be laws *beyond* physics; he believed that they would be part of physics, but new forms of physics, and that's what he hoped to find. Delbrück was somewhat disappointed when the structure of DNA came out. He was one of the first people to recognize how important it was—not everybody recognized that, by the way—but he thought it was too much like a tinkertoy. On the other hand, it was exactly the sort of thing that Pauling had been talking about; if it wasn't for a bit of bad luck, I think he would have got the structure of DNA himself.

So he had the right basic ideas—that chemistry and physical chemistry are important for biology and, as I've said, that the weak bonds (which were neglected at that time) are of extreme importance. The second one was the one I have already mentioned: the importance of shape and especially complementarity. But in addition to that, as we have seen, he initiated much fundamental experimental work. It wasn't that he just had ideas. He had a very active group, he taught a lot of people, but he also initiated a lot of research. The structure of small organic molecules was studied because he realized that by finding out the way small molecules packed together would give us enormous insight (as it has done) into big molecules (their size, shape and construction). In addition, as we have seen, he had several important detailed ideas: the alpha helix, the pleated sheet, the coiled coils, sickle cell as a molecular disease, the catalytic site of enzymes, and the molecular clock in evolution.

When I began to write this chapter, I realized that the title is not quite what it should be. "The Impact of Linus Pauling on Molecular

Biology" rather reads as if molecular biology existed to a considerable degree before Linus Pauling's influence, but that is not the case. The case really is that Linus Pauling was one of the founders, if not the major founder, of molecular biology. I'm distressed to notice that this isn't realized today by many of the young molecular biologists. Of course, it's often true that younger scientists are not greatly interested in the history of what went before them. They want to create history—not just to learn what happened before they came on the scene. It was because of Linus Pauling that our approach to the structural side of things had the character that it had, and it was successful because he had the right set of ideas. So we should salute Linus Pauling, not only for the wonderful things that he has done for chemistry and in particular his realizing the importance of quantum mechanics for chemistry and applying it as a chemist (as opposed to just a theorist)—but also for his absolute seminal role in getting molecular biology started.

5

How I Became Interested in the Chemical Bond: A Reminiscence

Linus Pauling
Linus Pauling Institute of Science and Medicine

My curiosity about the properties of substances developed early. When I was 12 years old I began reading about minerals. Agates were about the only minerals that I could collect, but I read books on mineralogy and copied tables of various properties of minerals. I wondered about hardness, streak, color, density, and crystalline face development, but it was only later that I began to get an understanding of the structural origin of the properties.

The next clear memory that I have along these lines is that when I was working as a paving plant inspector during the summer of 1919 in southern Oregon, living in a tent on the bank of Grave Creek, I spent much time checking over tables of properties of substances, such as the magnetic properties, in *The Handbook of Physics and Chemistry*. It is my memory that nothing much in the way of generalizations about properties in relation to composition came out of this effort. Then that fall I was appointed assistant instructor in quantitative chemical analysis in Oregon Agricultural College, where I worked full time during the academic year 1919–1920. At this time I read the 1916 paper

Figure 1. Linus Pauling at Oregon Agricultural College, age 20.

by G. N. Lewis on the shared-electron-pair chemical bond, and also several papers by Irving Langmuir, containing much greater detail than Lewis's paper. I was sufficiently interested to arrange that I would give a chemistry seminar that year. There were only two seminars given, one about the frozen fish industry by a man teaching agricultural chemistry and this one by me on the electron theory of the chemical bond.

In 1922 I had the remarkable good fortune to be appointed a graduate assistant in the California Institute of Technology and to have the recommendation made to me by Arthur Amos Noyes, chairman of the Division of Chemistry and Chemical Engineering, to carry on research in x-ray crystallography under the direction of Roscoe Gilkey Dickinson. I do not think that there was a better place in the world for me to go to prepare myself for my career than the California Institute of

FIGURE 2. A. A. Noyes driving his Cadillac (1917). From left to right: David F. Smith, Noyes, G. N. Lewis (on running board), Mrs. Lewis, Dr. James E. Bell, and Mrs. Bell.

FIGURE 3. Arthur Amos Noyes, 1917.

FIGURE 4. Roscoe Gilkey Dickinson.

Technology at that time, or that there was a better subject of research for me than x-ray crystallography.

Roscoe Dickinson had got his Ph.D. from C.I.T. in 1920, his dissertation being on the x-ray crystallography of sodium chlorate. He was in 1922 a National Research Council Fellow. Two pieces of x-ray apparatus were available, built by the instrument maker in the chemistry department of C.I.T. One, using radiation from an x-ray tube with a molybdenum anticathode, could, with an oscillating single crystal with a rather large developed face, give the photographic record of different orders of diffraction from that face, which then would serve for determining the length of the edges of the crystal unit of structure. The other piece of apparatus, producing a band of continuous radiation (wave lengths from about 0.24 Å up), used a thin stationary crystal to produce Laue photographs. The Laue photographic method was very

FIGURE 5. Sir Lawrence Bragg.

powerful, but for some reason was not used very much in laboratories other than at C.I.T.

I had read the book *X-Rays and Crystal Structure,* written by W. H. and W. L. Bragg, during the summer, and I had formulated a number of questions that I wanted to answer. At Oregon Agricultural College I had worked for two years as an assistant in mechanics and materials, including metallography, and I had become quite interested in metals and intermetallic compounds. I learned that no one had determined the structure of any intermetallic compound. Also, I was interested in learning the structures of various other inorganic substances. Only eight years had passed since the discovery of x-ray crystallography, and only a rather small number of structures had been determined, so that there was great opportunity for contributing to knowledge. During the first few weeks after my arrival in Pasadena near the end of

September 1922, Roscoe Dickinson taught me the elements of crystallography, and I began checking the literature, especially the several volumes of Groth's *Chemische Kristallographie,* looking for cubic crystals of substances with such composition as to interest me. In a couple of months I crystallized 14 substances. One, for example, was potassium nickel sulfate, anhydrous, $K_2Ni(SO_4)_2$. I dehydrated potassium sulfate and nickel sulfate, mixed the anhydrous powders, melted them in a crucible in a furnace that I myself had constructed, let the melt cool slowly, and obtained quite good crystals of the substance, 2 or 3 mm on edge, growing together but separable. Preliminary investigations with x-rays of these various crystals showed, however, that every one that I had selected was too complicated for its structure to be determined by the methods then available—there were too many parameters locating various atoms in the unit cell. After about two months Dickinson took me to the stockroom, got a nodule of molybdenite from the store of chemicals, cleaved the crystal into thin plates, and had me take rotation photographs and Laue photographs of it. Within a month the structure of this hexagonal crystal had been determined—rather interesting, because each molybdenum atom was surrounded by six atoms of sulfur at the corners of an equilateral prism, rather than of an octahedron, such as Dickinson had found in the potassium chlorostannate crystal whose structure he had determined.

I then determined the structure of an intermetallic compound, Mg_2Sn, and continued to investigate interesting substances by use of x-ray crystallography during the next three years.

At the same time I was serving as a teaching assistant in freshman chemistry, handling one of the sections of about a dozen undergraduate students, and was also taking a heavy load of courses, mainly in mathematics and theoretical physics. I was fortunate in that Richard Chace Tolman had come as professor of physical chemistry and mathematical physics to Caltech in 1921, and immediately began giving very interesting courses on quantum theory, atomic structure, the nature of science, and statistical mechanics. I was especially interested in statistical mechanics, and one of my first papers, written together with Tolman, was the application of statistical mechanics to a calculation of the residual entropy at low temperatures of a supercooled glass.

I was also interested in theoretical questions. I spent much time trying to develop an improved theory of the thermodynamic properties of

FIGURE 6. Richard Chance Tolman.

aqueous salt solutions, and I published one paper, together with Debye, in this field. I also was following with interest the debate between Millikan and other physicists on the one hand and G. N. Lewis on the other about the positions of electrons in molecules and crystals. Lewis argued that chemical phenomena showed that the electrons occupied definite fixed positions, such as at cube corners or tetrahedral corners, about the nucleus, whereas the physicists, who accepted the Bohr model, stated that the electrons were moving in orbits, and could not be at these fixed positions. In 1924 or 1925 I made models to illustrate my own ideas. I knew about Sommerfeld's idea that the electrons occupy elliptical orbits, and I assumed that in neon there are, in addition to the two inner electrons, four electron pairs occupying orbits of high eccentricity that are directed toward the four corners of a regular tetrahedron. I suggested also that methane had such a structure, and that accordingly both G. N. Lewis and the physicists were right, in that the electron pairs in the long orbits had average positions corresponding to G. N. Lewis's tetrahedron.

At this time, around 1924 and 1925, there was much discussion about the failure of the old quantum theory to give the right results in some cases, that is, results agreeing with experiment. It was found that sometimes introducing half-integral values of the quantum number caused the old quantum theory equations to come into agreement with experiment. I found an example of a similar difficulty. According to the old quantum theory, my calculations indicated that the dielectric constant of a gas such as HCl would be changed by the presence of a magnetic field. An experiment carried out in the physics department showed that this change did not occur, and later I was able to treat the problem with quantum mechanics and show that no change should occur according to the new theory. In the fall of 1925, after I had received my Ph.D. degree (in June) I was working as a National Research Council Fellow, but I applied for a European fellowship to the John Simon Guggenheim Memorial Foundation. Later the secretary of this Foundation, Henry Allen Moe, described my project as a study of the topology of the interior of the atom. I wrote to Niels Bohr and to Sommerfeld, asking permission to come to work in their Institutes. Bohr did not answer my letter, but Sommerfeld did. This was fortunate, because I am sure that I benefitted far more from Sommerfeld's lectures in Munich during the next year than I would have from working in the Bohr Institute in Copenhagen.

A. A. Noyes recommended that I resign my National Research Council fellowship and leave in February or March 1926 for Europe, supported for a couple of months by a grant from the California Institute of Technology, but with the expectation that I would receive the Guggenheim fellowship. Accordingly my wife and I set out for Munich, arriving there in April 1926. This was just at the time when Schrödinger was publishing his papers on the Schrödinger wave equation, and Sommerfeld began immediately lecturing about quantum mechanics, emphasizing the Schrödinger equation. During the same year, 1926, I was able to complete two papers on the application of quantum mechanics to the problem of the properties of atoms and monatomic ions. These two papers were among the earliest to be published in this field, aside from Schrödinger's. One of them dealt with the x-ray doublets and the other with the electric polarizability, diamagnetic susceptibility, and electron distribution in atoms and monatomic ions.

I was very fortunate to have gone to Sommerfeld's Institute of Theoretical Physics just at that time. For many years much of my work has consisted of the application of quantum mechanics to chemical problems. Both my experimental work, providing values of interatomic distances and bond angles, and my theoretical work have had significance for the question of the nature of the chemical bond.

My year in Munich was very productive. I not only got a very good grasp of quantum mechanics—by attending Sommerfeld's lectures on

FIGURE 7. Professor Arnold Sommerfeld, Ava Helen Pauling, and Linus Pauling, Jr., in 1928 in the California desert. Photo by Linus Pauling.

FIGURE 8. Linus Pauling in his laboratory, California Institute of Technology, circa 1951.

the subject, as well as other lectures by him and other people in the University, and also by my own study of published papers—but in addition I was able to begin attacking many problems dealing with the nature of the chemical bond by applying quantum mechanics to these problems.

I had, starting in 1922, been collecting the determined values of interatomic distances in crystals, and attempting to systematize them. I remember that in Munich there was some x-ray crystallography being carried out in Sommerfeld's Institute of Theoretical Physics. It was Sommerfeld's Privatdocent, Max von Laue, who discovered x-ray diffraction, and Sommerfeld had decided then, in 1913, to set up a small experimental unit in his Institute. One of the investigators there had determined a structure, that of zirconium disilicide, that I felt was surely wrong, because the interatomic distances, zirconium to silicon, were less than half what I thought they should be for a chemical bond between these atoms.

On my return to Pasadena I continued both the experimental work, mainly x-ray crystallography, and the theoretical work, getting what I felt was a better picture of the chemical bond. Finally, in 1939, I

FIGURE 9. Linus Pauling in the Concert Hall, Stockholm, about one hour after the Nobel Prize ceremony, 10 December 1954, with (from left to right) Mrs. Linus Pauling, Jr., Linda Pauling (daughter), and Ava Helen Pauling (wife).

collected information about the chemical bond and published the first edition of my book *The Nature of the Chemical Bond.*

There is one aspect of the chemical bond that seems to me still to be in an unsatisfactory state. This is the question of the nature of the chemical bonds in metals and intermetallic compounds. I am sure that there is still much room for progress in this field.

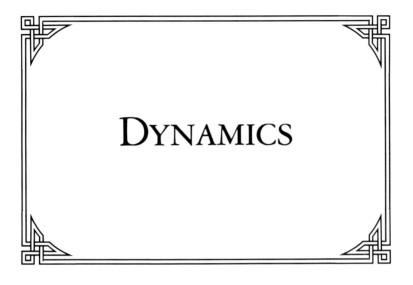

DYNAMICS

6

CHEMISTRY IN MICROTIME

Lord George Porter

Imperial College of Science, Technology and Medicine

All chemists of my generation were nurtured and inspired by Pauling's book *The Nature of the Chemical Bond,* which first appeared in 1938. This book which is dedicated to Linus Pauling is, therefore, appropriately entitled *The Chemical Bond: Structure and Dynamics.*

Chemistry, like most things, has static and dynamic aspects, dealing respectively with the structure of things and with how those structures change one into another. We may think of them as, respectively, chemistry in space, and chemistry in time. My assignment is with time.

I shall begin by looking at how, over the past century, we have come to understand quite a lot about chemical dynamics—about the rates and mechanisms of chemical change. This will lead naturally to shorter and shorter intervals of time as we probe the details of molecular motion. That is why this chapter is titled "Chemistry in Microtime" and why all four of the contributions about dynamics in this book will include a lot of fast talk.

Let us start at the beginning of the present century. The first Nobel prize for chemistry was awarded in the year Linus Pauling was born—1901—not to Linus himself (nine months was considered rather early even for him) but to J. H. Van't Hoff "in recognition of the extraordinary services he has rendered by the discovery of the laws of chemical dynamics. . . ." Van't Hoff had proposed, in 1884, several alternative equations for the rate of chemical reaction and Svante Arrhenius (1887) used one of these to propose what is universally known today as the Arrhenius equation:

$$k = Ze^{-E_a/RT}.$$

This could be given the simple interpretation, based on the earlier kinetic theory of gases of Maxwell and Boltzmann, that the rate of reaction, k, was equal to the number of collisions in unit time having, in two degrees of freedom, energy greater than the barrier to reaction E_a. In 1903, Arrhenius received the third Nobel prize for chemistry.

The experimental verification of theories of chemical kinetics was bedevilled by the complexity of what were apparently the simplest reactions. Most of these were gas-phase reactions, studied for example in the schools of Nernst (who recognized the first chain reaction, that between H_2 and Cl_2 with its photochemical quantum yields of up to a million), Bodenstein, and many others. It was not clear how collisional activation could lead to first-order kinetics and, for a short time, Perrin's radiation theory of activation received some support until F. A. Lindemann (1922; presented orally at a Faraday Discussion on Sept. 18, 1921) showed how the difficulty was removed if one expressed the reaction in two steps, one of activation and the second of reaction:

$$M + M \Leftrightarrow M^* + M,$$

$$M^* \Rightarrow Products.$$

This, and the Nernst chain reaction, were early indicators of the stepwise nature of chemical change and the importance of transient intermediates in the interpretation of the kinetics. However slow a chemical change may be, it will always involve, as intermediate steps, collisions occurring, at normal concentrations, at intervals of less than a billionth of a second.

But now Z in the Arrhenius equation was no longer remotely related to the collision number, except at very low pressures. A new concept was demanded—at least for unimolecular reactions.

TRANSITION STATE THEORY

Henry Eyring and Michael Polanyi, in 1931, developed a semi-empirical method, based on the Heitler–London theory, for constructing the potential-energy surface of a three-atom (of hydrogen) reaction. Although quantitatively suspect, these surfaces are qualitatively invaluable for envisioning the course of the reaction. Later Eyring (1935) and M. G. Evans and M. Polanyi (1935) independently developed what is now known as the "transition state" or "activated complex" theory. Here E_a in the Arrhenius equation is identified as the energy of the activated complex.

(When I was a young undergraduate, a new professor of physical chemistry, also very young, came to Leeds and gave the third-year lectures on quantum statistical thermodynamics and the transition state. His name was M. G. Evans—an inspiring lecturer with a passionate interest in chemical dynamics. It was "M. G." who made me a convert to physical chemistry. He succeeded Michael Polanyi in the Manchester chair but was to die of cancer only a few years later.)

Calculation of the pre-exponential factor, Z, began with the assumption that the transition state is in thermodynamic equilibrium with the reactants and the equilibrium constant is calculated in the usual way from the partition functions of the molecules concerned. One coordinate is assigned to the reaction path, and the forward rate coefficient becomes

$$k_r = (kT/h)K^{\#} = (kT/h)(Q^{\#}/Q_r)\exp(-E_a/kT),$$

where $K^{\#}$ is the equilibrium constant between reactants and transition state, Q_r is the product of partition functions (f) for the reactants, and $Q^{\#}$ is the similar product for the activated complex—less the partition function allocated to the reaction coordinate.

It is a reassuring part of this approach that the same universal constant kT/h is obtained whether one chooses translational, vibrational, or rotational partition functions for the reaction coordinate, and that the above equation leads directly to the collision theory expression between two atoms, with

$$Z = (kT/h)f_r^2 f_t^3,$$

where f_r and f_t are the partition functions for rotation and translation. But, of course, it takes us much further because the other molecular coordinates of rotation and vibration are now included via the Q terms. In its various refined forms, it is the basis of most of chemical kinetics still today. And the preexponential term, kT/h, which equals 6.25 × 10^{12} at 300 K, warns us that if ever we seek to observe this transition state we must be prepared to operate in times of ~10^{-13} seconds.

Just after the second world war there was an explosive growth in all branches of science, including the sciences of chemistry. Physical methods, and x-ray crystallography in particular, were making it possible to "see" atoms in their molecular context. Simultaneously, over the first half of the century, the science of molecular dynamics had developed rapidly but not in the equivalent sense—it was not yet possible to "see" those atoms in the course of chemical change.

By 1945, when I began research, the physical basis of molecular kinetics was fairly well advanced, if not amenable to exact computation. But it was quite rare to find an experimental rate which agreed even approximately with calculations based on the interaction of the initial reactants themselves. The new transition-state theories were capable of predicting reasonable values for the rates of elementary reactions, but very few chemical reactions were elementary. Most reactions of interest were sequential, often involving many steps. The lesson was learned, slowly and painfully, that the determination of mechanism was an essential preliminary to the application of absolute rate theory, or any other theory for that matter.

Intermediates in chemical reactions, such as those of combustion and polymerisation, were of great interest at the time, but the evidence for free atoms and radicals was all indirect and considered rather thin by chemists trained thoroughly in the conventional rules of chemical combination. For a time, therefore, it was considered quite important to provide evidence for the mere existence of these transient species.

It was usually accepted at this time that these intermediates could not, and probably never would, be studied directly during the course of the reaction. The situation is well illustrated in the introduction to the Faraday Society meeting on "The Labile Molecule" held in Oxford in 1947. No direct detection of free radicals is mentioned in the 400 pages of that discussion and, in his introduction, the President, H. W. Melville, wrote: "The direct physical methods of measurement simply cannot reach these magnitudes, far less make accurate measurements in a limited period of time, for example 10^{-3} sec." (Even the word millisecond had not entered the chemical vocabulary.) Indeed, no absorption spectrum of any radical other than diatomic was known and, of course, there was no spin resonance.

At the same meeting I read my first paper (with R. G. W. Norrish), on a study of the free radical methylene by the flow method of Paneth. I was very dissatisfied with it, and attempts using the same 7 kW mercury-arc source to observe the absorption spectrum of free radicals, such as methylene, in the steady state were also unsuccessful. Indeed no absorption spectra of polyatomic radicals had been observed at this time. The need for a more direct method was clear—preferably one which provided a time sequence of spectroscopic records.

So much has happened in chemical dynamics in the second half of this century that I must now focus on one small part of it. I shall devote the rest of my chapter to the development of methods for the study of very fast reactions and short lived intermediates and particularly the method called "flash photolysis," or "pulse-and-probe kinetic spectroscopy." Although dynamic studies in chemistry have, perhaps, not advanced science as spectacularly as have the revolutionary discoveries in biology arising from structural studies, the advance by nine orders of magnitude that has taken place in pulse/probe flash photolysis since the early 1960s is unmatched by any other technique of chemistry.

Other methods for the study of fast reactions, such as stopped-flow and temperature-jump methods, which are particularly valuable in solution, were developed over the same period, but none of them has a time resolution approaching that of flash photolysis. Then there are methods which sort out the molecules by velocity and orientation, and which provide far more detailed information than can be obtained from measurements only of average properties. These are discussed in following chapters by Professor Herschbach and by Professor John Polanyi, who has maintained so brilliantly the family tradition. I shall therefore now confine myself to flash photolysis.

The photographic recording of fast events was of course already well known in the last century. Fox Talbot used a spark to record a readable image of a rapidly spinning page of The Times at a Royal Institution discourse in 1851 and in California, appropriately enough, the first movie film was made by Eadweard Muybridge in 1887, for what must be the first study of fast biokinetics, since it established that, in a gallop, all four hooves of Governor Stanford's trotter left the ground.

But with molecules it is necessary to ride a number of horses at the same time, and, what is more, they must all run together in exact synchronism if one is to obtain a time sequence. This means initiating the reaction, as well as recording it, in a flash. The original flash photolysis apparatus, like the most modern ones, employed two flashes, the pump or photolysis flash to initiate change, and the probe or spectroscopic flash to record the situation a short time interval later. The principle is shown in Fig. 1. By repeating this at increasing time intervals a "movie" of the molecular change is recorded.

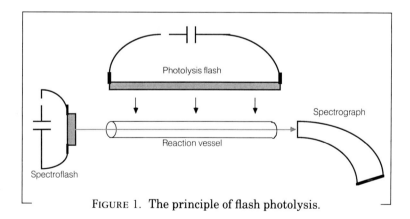

FIGURE 1. The principle of flash photolysis.

MICROSECOND ERA, 1950–6

The first apparatus, built in 1949, was large. Designed for gas kinetics, the reaction vessel was one metre long and there was a matching flash tube of the same length. As a consequence a high energy was used, and the 10,000 J flash lasted about one millisecond. A rotating spark gap (which was modelled on the SL radar) was used as shutter and trigger device. It soon became clear that energy could be sacrificed in favour of time resolution and, by using better capacitors, lower energy, electronic triggering, shorter reaction vessels, and shorter lamps, flashes and recordings of a few microseconds' duration were soon achieved. Attempts to use sparks to reduce the flash time further were not very successful, and the microsecond era of flash photolysis lasted just over 10 years.

The first results, submitted in 1949, included the observation of the absorption spectra of several diatomic radicals, ClO, CS, HS, the first polyatomic radical in absorption—HS_2—and the bleaching of chlorine molecules on dissociation into atoms. Later, many organic radical spectra were recorded for the first time. These included the aromatic radicals related to the original Gomberg free radical, triphenyl methyl, including

benzyl: $C_6H_5\,CH_2\cdot$, anilino: $C_6H_5\,NH\cdot$, phenoxyl: $C_6H_5\,O\cdot$,

their derivatives such as ketyls and semiquinones, the much more reactive phenyl radical $C_6H_5\cdot$, and many of its substituted derivatives. These were all observed in the gas phase, and the less reactive radicals, such as benzyl, were also recorded in solution. The absorption of phenyls is interesting in that it is assigned as a $\pi \rightarrow n$ transition—a new type. Smaller radicals of great chemical interest, particularly methyl and methylene, were observed and their spectra interpreted in detail by Herzberg and his group in Ottawa. To achieve this G. Herzberg and J. Shoosmith (1959) skillfully extended the technique into the far ultraviolet region.

I shall try to illustrate the conquest of the time barrier in experimental chemistry over the last 40 years with a few examples, old and new, that represent some of the principal areas of chemical dynamics.

In the microsecond regime we begin with three studies of the three main types of transient intermediate: atoms, free radicals, and excited states.

THE RECOMBINATION OF ATOMS

The three-body recombination of iodine atoms was one of the first flash kinetic studies carried out at Caltech by our Chairman, Norman Davidson and colleagues (1951) and by Ken Russell and John Simons in M. G. Evans' laboratory in Manchester, as well as in our laboratory in Cambridge. One would have thought that there was little room for kinetic complexity when two atoms such as iodine recombine in the presence of a third body M, but again such intermediates are involved. In this case no new spectra are observed and one merely follows the decreased density of I_2 absorption as a measure of iodine atom concentration.

It turned out that the "chaperone" effect (a term introduced after careful justification by N. K. Adam) involved the formation of a temporary complex between one atom and the chaperone:

$$I + M \Leftrightarrow IM \qquad\qquad k_1/k_2 = K$$

$$I + IM \Rightarrow I_2 + M \qquad\qquad k_3$$

The overall rate of recombination is then given by $k = k_3K$.
Evidence for this scheme included the following:

(1) The "activation energy," E_a, is negative with a value nearly equal to the heat of formation of IM.

(2) There is a variation of efficiency of chaperone molecules over a factor more than a thousand which it is difficult to account for on any other theory.

(3) There is a linear relation between the logarithm of the rate constant and E_a determined from the temperature coefficient of reaction rate (Fig. 2).

(4) One normally monitors the iodine atom concentration from the decrease in density of I_2 absorption, but with some of the more effi-

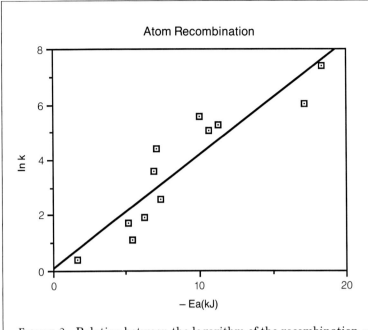

FIGURE 2. Relation between the logarithm of the recombination rate constant of iodine atoms and the measured "activation energy."

cient chaperones (e.g., CH_3I and NO), the species IM are observed as transients. The extent to which this mechanism applies also to cases of three atoms where M is a noble gas is a matter still under discussion at the present time.

THE CHLORINE–OXYGEN REACTION AND THE ClO RADICAL

ClO was the first radical to be studied in detail by direct kinetic observation and, although this took place in the early 1950s, very recent developments have put this molecule under further intensive study.

The spectrum attributed to ClO was obtained by flash photolysis of a mixture of chlorine and oxygen. It lived, under the conditions of the experiment, for a few thousandths of a second, and a time sequence of its absorption is shown in Fig. 3.

The main reactions were

$$Cl_2 + h\nu \Rightarrow 2\ Cl, \tag{1}$$

$$2\ Cl + O_2 \Rightarrow 2ClO, \tag{2}$$

$$2\ ClO \Rightarrow Cl_2 + O_2. \tag{3}$$

Consideration of the rates of these reactions made it necessary to introduce intermediate steps and transient molecules in reactions 2 and 3:

$$Cl + O_2 \Rightarrow ClOO$$

$$ClOO + Cl \Rightarrow 2ClO$$

$$2ClO \Leftrightarrow ClOOCl$$

$$ClOOCl \Rightarrow Cl_2 + O_2$$

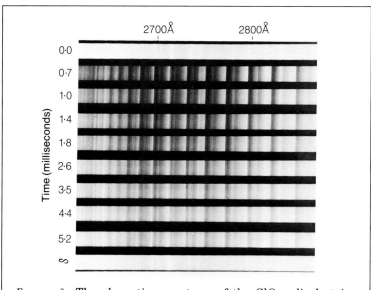

FIGURE 3. The absorption spectrum of the ClO radical at increasing time intervals after the flash photolysis of chlorine and oxygen.

And that was that. I was pleased with it, though when challenged I had to confess that I could not foresee any possible use of my work. I certainly did not foresee that this transient little molecule would one day be called the "smoking gun" which must be sought out and destroyed.

Later experiments were carried out on many aspects and variations of this problem. For example, ClO was prepared in other ways, and a particularly interesting observation was that, although the same chemical products, ClO and O_2, were formed from the two reactions

$$O + ClO_2 \Rightarrow ClO + O_2^*,$$

$$\text{and } Cl + O_3 \Rightarrow ClO^* + O_2,$$

most of the excess energy of reaction (more than 250 kJ/mol) is released as vibrational energy in the new bond that is formed—the oxygen molecule in the first case and the ClO molecule in the second case.

You will already have noted that the second reaction involves the destruction of ozone by chlorine. Clearly if there were any chlorine in the stratosphere this reaction would upset the steady-state concentration of ozone. But no sources of chlorine in the stratosphere were known so that, although the reactions of ClO continued to be of interest to photochemists through the sixties, they seemed to have little relevance to the ozone layer, or to anything else for that matter.

In 1970 James Lovelock found that traces of CFCs had pervaded the entire troposphere, but he saw no cause for alarm because they were unusually stable compounds. Fortunately, a Californian scientist, Professor F. Sherwood (Sherry) Rowland, began to wonder where, if they were so stable, they finished up and concluded that the answer was the stratosphere where they would be broken down by the ultraviolet light to give chlorine which in turn would catalyse the decomposition of ozone.

The saga of what followed is well known and the matter was contentious until just four years ago when, in August and September 1987, a large airborne American experimental group overflew Antarctica and measured the concentrations of several trace gases up to altitudes of 18 km. The results were as convincing as we are ever likely to be blessed with.

The free radical, ClO, appeared rapidly and abundantly in late August, and this was followed by a dramatic fall of ozone in mid-September, which showed an inverse correlation even in the smallest details (see Fig. 4). All other gases, such as the nitrogen oxides and even water, decreased with ozone; only ClO increased. The buildup of ClO during the winter was due to the stability of the ClOOCl dimer at low temperatures in the dark. The "smoking gun" had at last been found and all the fingerprints were in place. In that month the Montreal protocol was signed by 24 countries and later meetings were attended by 123 countries.

I like to tell this story as an illustration of what we scientists know very well—though unfortunately our practical-minded paymasters often seem to be unaware of it—new scientific knowledge, however useless it may seem at the time, is not only likely to be exploited one day but it is wise to have it ready to cope with unforeseen problems or disasters.

ELECTRONICALLY EXCITED STATES

The most important postwar development in photochemistry has been the recognition that the electronically excited state is a distinct substance with its own structure and reactivity. Light is no longer regarded as a mere activator, a sort of Bunsen burner, and the excited state is no longer just a hot molecule—it is a new chemical species. The next step was clearly to detect these excited states directly, by time-resolved absorption spectroscopy, but the initial singlet state reached by absorption was far too short-lived to be observable.

In the mid-1940s G. N. Lewis with Michael Kasha (Lewis and Kasha, 1944) and other colleagues at Berkeley, and also Terenin in Leningrad, showed that molecular phosphorescence is emission from the triplet state which can be remarkably stable in rigid solutions, sometimes with lifetimes as long as a second. Following a paper by Donald McClure (1951), who detected triplet state *absorption* in rigid glasses, I asked a new student, Maurice Windsor, to try to detect the triplet states of aromatic hydrocarbons in ordinary fluid solutions. It was his first task; he was successful, and one of his kinetic records of

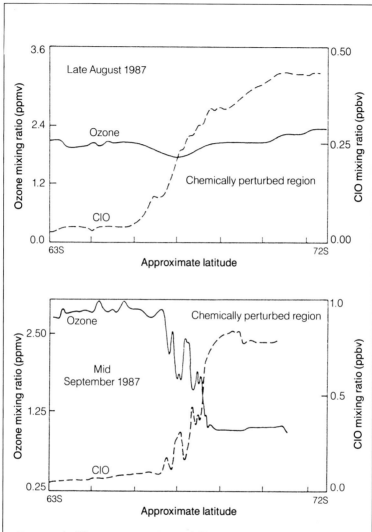

FIGURE 4. The concentrations of ClO and ozone over Antartica in August and September 1987, showing the close inverse correlation between the two gases (measurements by J. G. Anderson (ClO), and by M.H. Proffitt and W.L. Starr (ozone)).

the triplet states of aromatic molecules in solution is shown in Fig. 5. Almost immediately, another student, Franklin Wright, observed the same triplets in the gas phase.

The long life of triplet states (typically in the millisecond/microsecond range for organic molecules in solution), which results from the spin-forbidden nature of their transition to the ground state, makes them responsible for most of the photochemical changes that occur in solutions and gases (but not necessarily in biological systems, as we

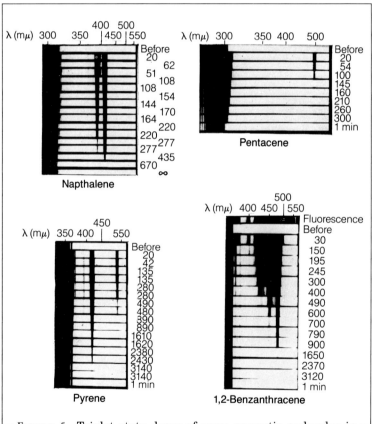

FIGURE 5. Triplet state decay of some aromatic molecules in solution.

shall see later). It also makes triplets easy subjects for kinetic study by flash photolysis, and measurements of their chemical reactivity and their physicochemical properties, such as pKs and dipole moments, are now routine.

Carbonyl compounds have always been of particular interest to photochemists, and the first direct absorption study of reacting excited states was on the reactions of quinones and ketones with organic solvents where both the triplet-state absorption and the semiquinone radicals and radical ions were readily observed and measured.

Here again there was a practical application of the work because the phototendering of fabrics coloured with anthraquinone and similar dyes was a serious problem. A single substituent such as an amino group will "protect" the fabric with the result that blue dyes are generally less reactive than red ones—a fact long known but without any rational explanation. The general reaction of carbonyl compounds with organic materials is

$$>C{=}O + RH \rightarrow)C'{-}OH + R.$$

It occurs from the triplet state, but the reactivity varies from zero to nearly unit efficiency from one quinone to another. There are several reasons for this—reversible intramolecular reaction from a group *ortho* to the carbonyl is one of them—but there is a more general and subtle reason. This results from the fact that the excited states of carbonyl derivatives, whether singlet or triplet, may again be classified two types, designated $\pi{-}\pi^*$ and $n{-}\pi^*$ by Kasha in 1950.

In an $n{-}\pi^*$ state the electron configuration is one where a nonbonding electron has been transferred onto the oxygen of the carbonyl group, whereas the opposite situation is the case for the $\pi{-}\pi^*$ state in most quinones and aromatic ketones which have a strong dipole corresponding to intramolecular electron transfer *from* the carbonyl group to the rest of the molecule. Hydrogen atom transfer to the carbonyl, like electron transfer, requires an electrophilic oxygen so that the reactivity for hydrogen abstraction by carbonyl is enhanced over that of the ground state in an $n{-}\pi^*$ state but may be actually reduced in a $\pi{-}\pi^*$ state, especially one with strong charge-transfer character.

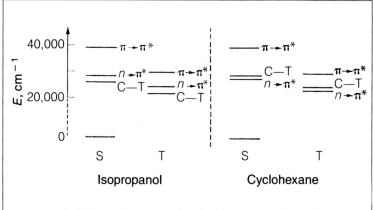

FIGURE 6. Electronic energy levels of p-amino benzophenone, showing how the $n-\pi^*$ state is the lower only in nonpolar solvents.

But now the situation becomes more interesting still when we consider the effect of polar and nonpolar solvents on the energies of the two states with their very different dipole moments.

In a molecule like p-amino benzophenone, the $\pi-\pi^*$ state has much charge-transfer character and is more stable, and therefore of lower energy, than the $n-\pi^*$ state in polar solvents, whereas the $n-\pi^*$ state was found to be the lower in hydrocarbon solvents (see Fig. 6).

We therefore predicted that this molecule would abstract hydrogen atoms from hydrocarbons but not from ethanol (Porter and Suppan, 1964). This turns ground-state chemistry on its head, but that is exactly what was found to happen.

In these and similar studies, it was now almost as easy to follow the dynamics of excited states, and the free radicals formed from them, as to time reactions with a stopwatch. But many events and species of great interest, including the first excited singlet states, were now clearly too fast for microsecond techniques. Fast oscilloscopes and detectors (streak cameras) were being developed, but the only submicrosecond flashes were sparks which, although satisfactory for single-photon-counting fluorescence kinetics, did not have enough energy for flash photolysis. A new source of short high-energy light pulses was needed, and the laser appeared just on time.

NANOSECOND ERA, 1960–

The laser was discovered in 1960, but it was several years later that it was applied to photochemistry, and the time resolution of flash techniques stuck round about a few microseconds until the mid-1960s. The first ruby-laser pulses, activated by a conventional flash lamp, had the same duration as the flash, which was many microseconds, but nanosecond exciting flashes of adequate power soon became available from Q-switched lasers and were used for the detection of a multiphoton photochemical process in 1966 (Porter and Steinfeld, 1966). One of the difficulties was the absence of a suitable probe flash; continuous monitoring sources were available, but there were no suitable submicrosecond flash sources of white light or tunable monochromatic sources.

The pulse/probe principle, shown in Fig. 7, which was to become obligatory when still shorter pulses became available, was used in

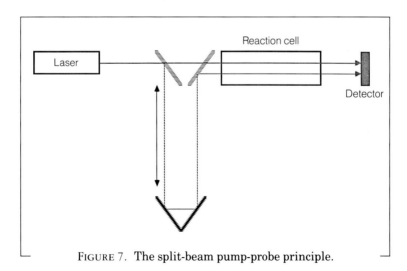

FIGURE 7. The split-beam pump-probe principle.

1968 (Porter and Topp). In this apparatus, a diagram of which is shown in Fig. 8,

(1) A single flash split into two beams was used because precise synchronisation of two separate flashes became increasingly difficult.

(2) A white light continuum of the probe pulse was generated as the probe beam—in this case by exciting a broad-band fluorescencing dye with the monochromatic flash.

(3) The delay between pulse and probe was provided by a variable optical path in the probe beam.

The excited singlet-state absorption was immediately added to the list of transient species amenable to direct flash photolysis study (see

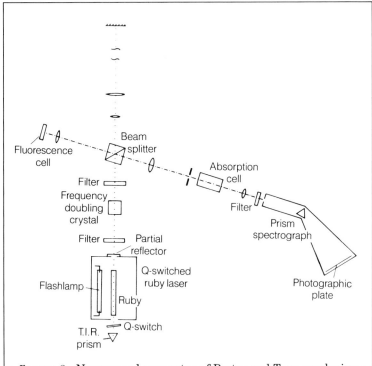

FIGURE 8. Nanosecond apparatus of Porter and Topp employing a split beam with a fluorescent "white" probe.

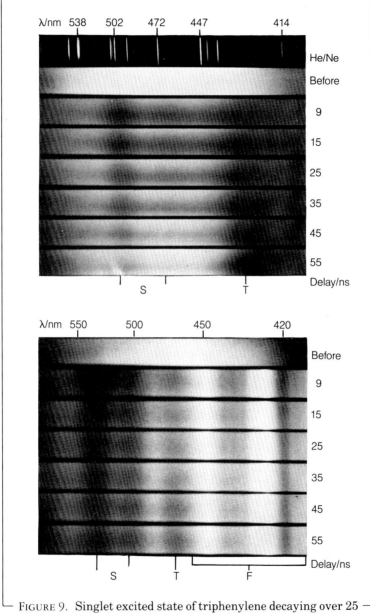

FIGURE 9. Singlet excited state of triphenylene decaying over 25 ns to give a triplet.

Fig. 9). New kinds of transients also became known, many of them of great significance in the understanding of photochemical mechanisms, transients such as excimers and exciplexes, intermolecular charge-transfer complexes, and their separated ions. Intermolecular electronic energy transfer could now be studied directly between triplet as well as singlet states; its rate at distances not involving orbital overlap was found to be well described by the inductive resonance theory of Förster.

The radiationless conversion processes which were at the heart of intersystem crossing to and from triplet states were something of a mystery at first. It became apparent that they, as well as intermolecular energy transfer, molecular dissociation, and electron transfer between molecules, were formally similar and were to be interpreted in terms of the quantum states of the system as a whole (involving both molecules in the intermolecular case). The rate of transition from one partial state to another is given in every case by expressions based on the Fermi golden rule:

$$k_{et} = 4\pi^2/h \mid V(r) \mid^2 FC,$$

where $V(r)$ is the electronic coupling of the reaction and product states and FC, the Franck–Condon factor, describes the overlap of vibrational wave functions between these states for which an exact quantum mechanical expression (assuming harmonic nuclear motions) has been given by Marcus and Sutin (1985).

PICOSECOND ERA, 1970–

The nanosecond era, like all these time regions, will always provide interesting problems of research, but it was only a short time before faster pulses became available. Further developments in time resolution of flash photolysis now depended almost entirely on what the laser physicists could provide. As soon as mode-locked lasers became available, measurements were extended into the picosecond region. The split-beam pump-probe principle now became essential.

The one new development that was now necessary, if the advantage of full-band monitoring was to be retained, was the generation of white-light picosecond pulses from the narrow-band pulses of the laser. The discovery was made by Alfano and Shapiro that such pulses can be generated simply by focussing the beam into a medium such as water. This occurs, without extending the pulse duration, by self-phase-modulation of high-intensity "monochromatic" pulses. It was an important step because it solved the continuum flash problem for all times, making possible "conventional" flash photolysis, with a monochromatic pulse flash and white-light probe flash down to the limits of pulse duration, however short. The numerous studies on this time scale, many of them in solutions and in biological systems, include orientational and vibrational relaxation, energy transfer, and electron transfer.

Femtosecond Era, 1980–

The extension of time resolution into the femtosecond range resulted largely from two improvements, the CPM laser and pulse compression, derived from Shank and his colleagues at Bell labs who eventually measured pulses as short as 6 femtoseconds.

To obtain useful information from measurements on such short time scales, it is essential to have high sensitivity and reproducibility as well as short pulses. Two typical state-of-the-art femtosecond systems for studies in chemistry are those in operation at Caltech and at Imperial College, London. The apparatus designed by David Klug, James Durant, and their colleagues in our laboratory at Imperial College is shown in Fig. 10. The 100 fs pulse train is generated in a colliding pulse-mode laser and amplified in a bow-tie dye jet through which the beam makes six passes. The whole amplified pulse is then focussed into a water cell where it is converted into white-light pulses of the same duration. The beam is now split into two parts; the first is filtered and then amplified in a second bow-tie jet arrangement so as to give a narrow exciting pulse of any chosen wavelength, and the second becomes the probe pulse of white light. The pulse repetition rate of 60 kHz,

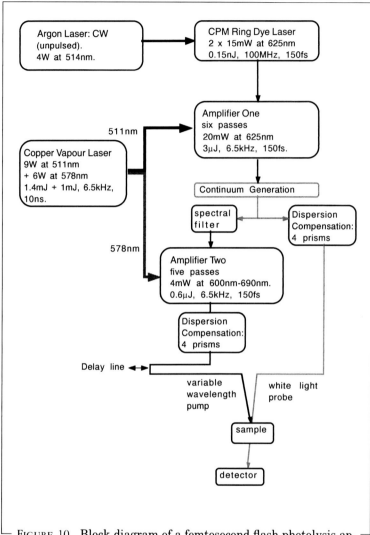

FIGURE 10. Block diagram of a femtosecond flash photolysis apparatus (D. Klug, J. Durant, *et al.*, Imperial College, London).

made possible by the copper vapour laser, allows detection, over noise, in a few minutes' signal averaging, of absorption transients having an optical density of 0.001, and with time resolution of 100 fs.

I shall conclude by describing the study of two very important but very different subpicosecond reactions that lie at the two extremes of molecular complexity. The first is the passage of a simple molecule through the transition state; the second is the passage of an electron through the reaction centre of the photosynthetic unit.

FEMTOSECOND SPECTROSCOPY OF THE TRANSITION STATE

The Z factor of $kT/h \sim 10^{13}$ s seems to be the upper limit for rates of chemical change where nuclei, rather than just electrons or energy, change their position. It corresponds to a minimum lifetime of ~ 100 fs for the transition state, so with femtosecond resolution it should be possible, in principle, to follow the fastest reactions as they proceed through the activation barrier.

Ahmed Zewail, at Caltech, has been able to do just this and, with his permission, I should like to describe two of his elegant experiments. (It is worth noting that these experiments were carried out in the room that once housed Linus Pauling's x-ray machines.)

DISSOCIATION IN THE EXCITED STATE

Here molecular excitation occurs to an upper dissociative level and as the bond breaks and the molecular fragments run down the upper potential-energy surface, the probe flash monitors either the reactant, i.e., the dissociating molecule, or the products, as they are formed (see Fig. 11).

The ICN molecule is excited by a 60 femtosecond pump pulse at 307 nm which raises it to the upper curve from which it begins to dissociate:

$$\text{ICN} + h\nu \rightarrow \text{ICN}^* \rightarrow \text{I}\cdots\text{CN} \rightarrow \text{I} + \text{CN}.$$

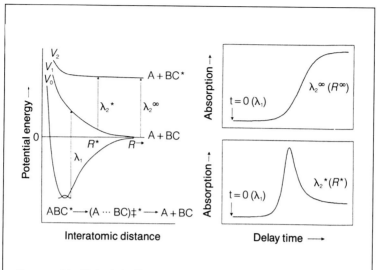

FIGURE 11. Principle for a spectroscopic study of the transition state in a dissociating ABC molecule. The potential energy diagram on the left shows how the molecule traversing the repulsive upper curve, reached by absorption at λ_1, is monitored by absorption at λ_2 up to $\lambda_2 \infty$, and the curves on the right show the predicted kinetics at two wavelengths. (From A. H. Zewail, *Science* **242**, 1645–1653. Copyright © 1988 by the AAAS.)

The final product CN is monitored by its fluorescence induced by a probe pulse of 388.9 nm. Intermediate points, taken whilst the bond is only partially separated, are monitored by fluorescence which is excited by probe pulses of slightly longer wavelength:

$$CN + h\nu \ (388.9 \ nm) \rightarrow CN^* \rightarrow fluorescence + CN,$$

$$I \cdots CN + h\nu \ (389.5 \ nm) \rightarrow I \cdots CN^* \rightarrow fluorescence + I \cdots CN.$$

The CN fragment grows monotonically to a maximum (see Fig. 12, upper curve), whilst the intermediate $I \cdots CN$ grows to a maximum as the bond expands and then decays again as the bond expands further and dissociates (see Fig. 12, lower curve). The recoil velocity of fragments is typically 1 km/second, or 1 Å/100 femtoseconds, and the decay time is 205 ± 30 fs (Fig. 12).

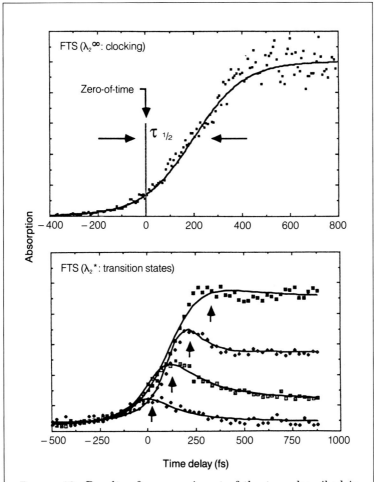

FIGURE 12. Results of an experiment of the type described in Fig. 11 for the dissociation of the molecule ICN. The transients are taken for $\lambda_2 \infty$ 388.5 nm and for λ_2 = 389.7, 389.8, 390.4, and 391.4 nm. The detection is by fluorescence of CN*. (From A. H. Zewail, *Science* **242**, 1645–1653. Copyright © 1988 by the AAAS.)

DISSOCIATION OF SODIUM IODIDE

The second example from Zewail is the dissociation of sodium iodide. The method is similar, but now there are two intersecting upper surfaces, the lower dissociating to Na and I atoms, and the higher to Na^+ and I^-, as shown in the top part of Fig. 13. If the excited molecule crosses to the ionic surface, it becomes bound and may oscillate back and forth until eventually it crosses, with a probability that is found to be 0.1, to the dissociative lower covalent surface.

The molecule is excited by light of wavelength 310 nm, and if the fluorescence of the free sodium atom is monitored, the upper curve of the lower half of Fig. 13 is obtained. If off-resonance is monitored at 589 nm the intermediate transition state trapped in the upper well is observed as shown in the lower curve. The round-trip time in the well is seen to be ~1 ps. In other examples, it was possible to follow the transition state over the actual saddle point.

Here we see a series of records, taken in real time, of the fastest of all steps in a chemical change. Other beautiful methods of exploring the transition states of molecules and the theory behind them will be described in detail in following chapters by John Polanyi and Dudley Herschbach.

ENERGY AND ELECTRON TRANSFER IN PHOTOSYNTHESIS

I now turn to the dynamics of very fast change in some of the most complex molecular systems whose reactions have been observed in real time.

The study of photosynthesis is, and probably always will be, the greatest odyssey of the photochemist. At the present time, it is at a very exciting stage for two reasons. First, techniques for the isolation, crystallisation, and structure determination of the pigment–protein complexes of photosynthetic organisms are advancing rapidly. Second, the primary photochemical processes of energy transfer, and some of those of electron transfer, occur near the limits of time resolution, even of present techniques. Since ordinary chemical processes in this time

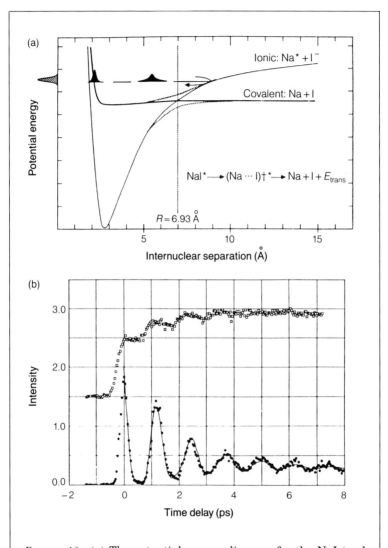

FIGURE 13. (a) The potential energy diagram for the NaI (and Na⁺I⁻) molecules. (b) The observed fluorescence of Na showing traversals within the potential energy well. Approximately 10% dissociation occurs on each traversal. The molecular excitation was at 310 nm, the upper curve was obtained by exciting Na on-resonance and the lower curve by exciting off-resonance at 589 nm. (From A. H. Zewail, *Science* **242**, 1645–1653. Copyright © 1988 by the AAAS.)

range are rare, chemists who like to work fast are turning to these biological problems with enthusiasm.

The pigment–protein complexes in the photosynthetic membranes of plants and bacteria perform two principal functions—light harvesting in the antennae and electron/proton transport in the reaction centres.

Most of the chlorophyll and most of its fluorescence are associated with the light-harvesting antennae, where there may be 300 times more chlorophyll than in the reaction centres. The lifetime of most of the excited singlet chlorophyll measured in the whole chloroplast is ~400 ps, which is interpreted as the average time taken for the excitation to be trapped at the reaction centre. This is in fair agreement with calculations of random-walk energy transfers among 300 molecules by a Förster-type inductive resonance mechanism, each individual step transfer of which occurs in a time of a few hundred femtoseconds.

Observations of energy transfer in the picosecond region have been mainly carried out by fluorescence lifetime measurements, usually with single photon counting, which is technically easier than transient absorption techniques. One of the early successes in this area was the observation, using streak camera detection, of the successive energy transfer between four different pigment layers in the antennae phycobilisomes of red and blue-green algae (Porter *et al.,* 1978).

The energy transfer sequence in the red alga *Porphyidium cruentum,* with associated times for peak excitation, was as follows:

$$\text{phycoerythrin} + h\nu\ (530\ \text{nm}) \overset{0\ \text{ps}}{\rightarrow} \text{phycoerythrin}^* \overset{13\ \text{ps}}{\rightarrow} \text{phycocyanin}^*$$

$$\overset{24\ \text{ps}}{\rightarrow} \text{allophycocyanin}^* \rightarrow \text{chlorophyll}^*,$$

50 ps lifetime in whole algae, 4,000 ps in isolated phycobilisomes.

The ratio of these times to the lifetime of the excited states of the isolated pigments shows that these energy transfers are more than 99% efficient.

Over the last few years those who are able to apply pico- and sub-picosecond techniques to the problems of photosynthesis have left energy transfer, in spite of the many problems that remain, and have concentrated on the even more interesting *electron*-transfer processes of the reaction centre. Most of these have used photosynthetic bacteria (*Rhodopseudomonas viridis* and *Rhodospirilum rubrum*) which possess only one reaction centre and are far simpler than the chloroplast of the green plant.

REACTION CENTRES OF PHOTOSYNTHETIC BACTERIA

The reaction centre of *Rhodopseudomonas viridis* was first isolated in 1978 (Clayton and Clayton, 1978). It was shown to have four molecules of bacteriochlorophyll (BChl), two of which were probably a dimer, two molecules of bacteriopheophytin (BPh), two or more ubiquinones (UQ), and a non-haem iron. Flash photolysis studies in the picosecond region were carried out by Rockley *et al.* (1975); by Kaufmann *et al.* (1975); and by Holten *et al.* (1978), who established that electron transfer probably occurs in the expected sequence of the known energies of these pigments and led to predictions of the sequence and times of transfer as follows:

$$(BCl)_2^* \cdot\cdot BPh \xrightarrow{4 \text{ ps}} (BCl)_2^+ \cdot\cdot BPh^-,$$

$$BPh^- \cdots UQ \xrightarrow{200 \text{ ps}} BPh \cdots UQ^-.$$

In 1981 Hartmut Michel in Martinsried reported the successful crystallisation of the photosynthetic reaction centre from *Rhodopseudomonas viridis*. Six years later, with the collaboration of Johan Deisenhofer, the structure of the reaction centre had been determined to 2.3 Å resolution. The structure confirmed the order of the electron transfer molecular sequence as it had been derived from flash photolysis, and the separations were consistent with the measured electron transfer times, provided the BChl, which bridges the gap of 17 Å between $(BCl)_2$ and BPh, mediates in the first electron transfer step above with a sub-picosecond electron transfer from BCh to BPh.

This tour de force of Michel, Huber, and Deisenhofer, following the dynamic studies, provided a perfect marriage of "chemical structure and dynamics" that is the theme of this book. It was another step forward in the x-ray crystallographic determination of structures of ever-increasing complexity, and it confirmed in a dramatic way the predictions about the electron transport sequence which had been made on the basis of kinetic studies. Above all, the heart of the photosynthetic process was now clearly revealed. Shortly afterwards the structure of another reaction centre, of a similar bacterium, *Rhodobacter sphaeroides,* was elucidated in a similar way, and a very similar structure of the electron transfer sequence was found (Allen *et al.,* 1987).

There are few, if any, chemical systems involving a sequence of electron transfers whose structure and kinetics are known in the detail of those of the photosynthetic reaction centre, and these new discoveries have therefore proved to be useful as a model for testing theories of electron transfer.

As discussed earlier, most calculations of the rate of electron transfer are based on the Fermi golden rule. In this expression the electronic matrix element for coupling between the two interacting species, $|V(r)|$, is taken to decrease exponentially with the edge-to-edge separation between them so that

$$k_r = k_{r.vdw} e^{-\beta(r - r.vdw)},$$

where $k_{r.vdw}$ is the rate of electron transfer at the minimum van der Waals distance.

It seems that in the photosynthetic bacteria the rate comes to depend principally on the distance-dependent electronic function $V(r)$, and that differences in the Franck–Condon factors are less important. A linear relation is therefore found between ln $k(r)$ and the separation, r, as was shown to be the case by Dutton *et al.* for 15 electron transfers in the two structurally known photosynthetic bacteria, with the van der Waals distance $r_{vdw} = 3.6$ Å, $\beta = 1.35$ Å, and $k_{r.vdw} = 10^{13}$ s^{-1} (Moser *et al.,* 1990) (Fig. 14).

The protein seems to be merely a scaffolding which determines the appropriate distance for electron transfer through what is essentially a homogeneous solvent.

REACTION CENTRES OF THE CHLOROPLAST

We are beginning to turn now to the reaction centres of the higher plants which, although of more practical importance, are more compli-

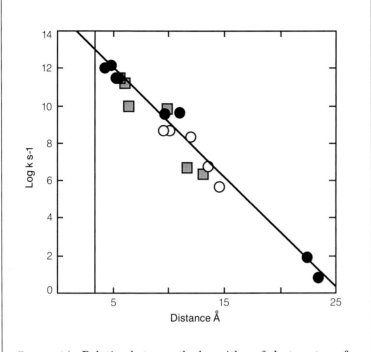

FIGURE 14. Relation between the logarithm of electron-transfer rate and the separation r in photosynthetic bacteria. (From Moser, *et al.* (1990), Electron and Proton transfer in chemistry and biology. Proceedings of the Bielefeld workshop, with permission.)

cated and poorly characterised. It is well established, however, that there are two separate photosystems, PS1 and PS2, in the chloroplast of the green plant, each with its own reaction centre and light-harvesting units. Two-dimensional crystals and a structure at 17 Å resolution have been obtained by Kühlbrandt for the light-harvesting antennae of photosystem 2 but not for the reaction centres. There has, however, been good progress recently in the isolation of the reaction centres of photosystem 1 and photosystem 2, and flash photolysis studies have begun on these centres in the picosecond and femtosecond regions.

Photosystem 1

Preparations of the reaction centre of PS 1 available at present re-
tain a small antenna unit of about 30 chlorophyll molecules. Although
this is a disadvantage for the studies of the electron transport sequence, it provides a useful system for the study of energy transfer in
the antenna followed by the first electron transfer step. In Fig. 15, the
bleaching of the broad absorption band of the antenna chlorophyll
around 690 nm is seen immediately after the flash. As this absorption
decays with a lifetime of 15 ± 3 ps, simultaneously a new sharp absorption band appears at 700 nm, which is the result of energy transfer
from the antenna chlorophyll to P700 followed by a rapid electron
transfer to pheophytin (Pheo) to form $P700^+–Pheo^-$. The full curves
are the spectra observed with reaction centres in which P700 had been
chemically reduced, whilst in the broken curves it had been previously
oxidised. The absence of the transient due to P700 oxidation in the
oxidised case is as expected, but the almost identical kinetics of the
return to the ground state is surprising and indicates that the antenna
excitation is quenched by the oxidised P700 as efficiently as by the
reduced form. The residual bleaching of the antenna chlorophyll in
both cases at 270 ps is attributed to formation of the long-lived triplet
state (Gore *et al.*, 1986).

Photosystem 2

This is of the greatest interest because unlike PS1 or the bacteria, it
is able to librate oxygen from water. Furthermore, its protein shows
considerable homology with that of the photosynthetic bacteria, and
its pigment electron transfer sequence also shows striking similarities.
The reaction centre, free of antenna chlorophyll, has recently been isolated by Satoh and by Barber, and the transients following absorption
by P680 of a 100 fs pulse have been studied by David Klug, James
Durant, Linda Giorgi, James Barber, and their colleagues at Imperial
College. The deactivation of the excited state, observed as the decay of
its stimulated emission, and identified with the first electron transfer,
is found to occur in 1.1 picoseconds, somewhat faster than the equivalent process in photosynthetic bacteria (Fig. 16).

These photosynthetic systems are impressive in their efficiency and

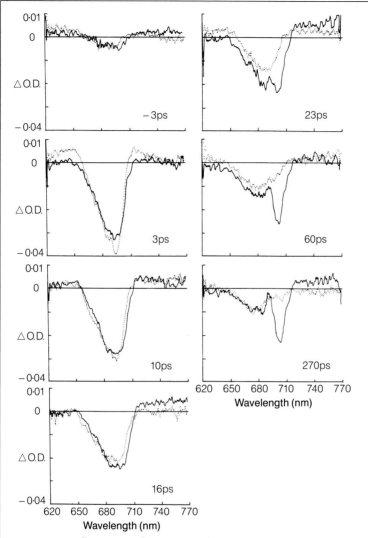

FIGURE 15. Sequence of transient absorptions in PS 1 reaction centres from pea chloroplasts showing energy transfer with a time constant of 15 ± 1 ps (from the small antenna system absorbing at to P700) to P700 and oxidation of the latter. ___ is for P700 chemically reduced and ...... is for P700 chemically oxidised.

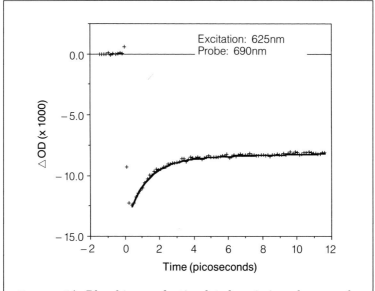

FIGURE 16. Bleaching and stimulated emission changes observed at 690 nm following excitation of isolated P.S.2 D1/D2 reaction centre particles. The maximum optical density change is $- 1.4 \times 10^{-2}$. The excitation pulse had an energy of 1μJ and duration 150fs. The solid curve shows the best fit to biexponential decay with lifetimes of 700fs and 30ps.

complexity and one asks of Nature, with the greatest respect, how and why it does things the strange way it does. What, for example, is the purpose of having four or more different pigment molecules packaged into a sequence across the membrane. And how is the back reaction prevented?

The answer to the second question is probably given in the first. The reverse reaction (that going by the reverse path of the forward reaction and therefore returning to the excited state) is prevented on purely energetic (endergonic) grounds. The back reaction (that proceeding to ground-state products and therefore energetically downhill) is prevented by the spacial separation of charge which occurs when the electron has moved from one end to the other of the electron transport chain; this creates a high activation-energy barrier to tunneling, even though the reaction overall is exergonic. This seems to be the reason, probably the only reason, why Nature effects its electron transport processes indirectly via a number of intermediate redox couples which separate the initial donor from the final acceptor.

Nature has evolved processes which are beyond the abilities of the chemist to reproduce in his laboratory even when the principles are understood. Nevertheless, one of the great achievements which we may anticipate during the next decade will be the elucidation of the chemical dynamics of the photosynthetic processes of the green leaf.

In chemical dynamics, having followed reactions through their transition state, what else may we expect in the next few years? A further improvement in the time resolution of flash photolysis by another factor of a thousand or so into the attosecond region? Not in photochemical studies, unless x-rays are included, because such time resolution is not possible in chemistry within the restrictions of the uncertainty principle, nor, for the same reason, is it useful. The uncertainty in energy associated with a measurement made in one femtosecond is comparable with chemical bond energies, and a pulse of one femtosecond in the red region of the spectrum contains only two wavelengths. A little qualitative information may be derived beyond this, but then chemistry, as a measurable science, comes to an end.

But we have the sure consolation that there will always be plenty of novel and fascinating molecular changes for chemists to explore in the seconds, milliseconds, microseconds, nanoseconds, picoseconds, and femtoseconds of their lifetimes.

REFERENCES

Alfano, R. R., and Shapiro, S. L. (1970). *Phys. Rev. Lett.* **24,** 592.

Allen, J. P., Feher, G., Yeats, T. O., Komiya, H., and Rees, D. C. (1987). *Proc. Nat. Acad. Sci. USA* **84,** 5730–5734.

Arrhenius, S. (1887). *Z. Physik. Chem.* **1,** 110.

Clayton, R. K., and Clayton, B. J. (1978). *Biochim. Biophys. Acta* **501,** 478–487.

Davidson, N., Marshall, R., Larsh, A. E., and Carrington, T. (1951). *J. Chem. Phys.* **19,** 1311.

Evans, M. G., and Polanyi, M. (1935). *Trans. Faraday Soc.* **31,** 875–894.

Eyring, H. (1935). *J. Chem. Phys.* **3,** 107–115.

Eyring, H., and Polanyi, M. (1931). *Z. Physic. Chem.* **B12,** 279–311.

Gore, B. L., Doust, T. M., Giorgi, L. B., Klug, D. R., Ide, J. P., Crystall, B., and Porter, G. (1986). *J. Chem. Soc. Faraday Trans. 2* **82,** 2111–2115.

Herzberg, G., and Shoosmith, J. (1959). *Nature* **183,** 1801.

Holten, D., Windsor, M. W., Parson, W. W., and Thornber, J. P. (1978). *Biochim. Biophys. Acta* **501,** 112–126.

Kaufmann, K. J., Dutton, P. L., Netzel, T. L., Leigh, J. S., and Rentzepis, P. M. (1975). *Science* **188,** 1301–1304.

Lewis, G. N., and Kasha, M. (1944). *J. Amer. Chem. Soc.* **66,** 2100.

Lindemann, F. A. (1922). *Trans. Faraday Soc.* **17,** 598–599.

Marcus, R. A., and Sutin, N. (1985). *Biochim. Biophys. Acta* **811,** 265–322.

McClure, D. (1951). *J. Chem. Phys.* **19,** 670, 185.

Moser, C. C., Keske, J. M., Warncke, K., and Dutton, P. L. (1990). Electron and proton transfer in chemistry and biology. Proceedings of the Bielefeld workshop.

Porter, G., and Steinfeld, J. I. (1966). *J. Chem. Phys.* **45,** 3456.

Porter, G., and Suppan, P. (1964). *Proc. Chem. Soc.,* 191.

Porter, G., Tredwell, C. J., Searle, G. F. W., and Barber, J. (1978). *Biochim. Biophys. Acta* **501,** 232–246.

Rockley, M. G., Windsor, M. W., Cogdell, R. J., and Parson, W. W. (1975). *Proc. Nat. Acad. Sci. USA* **72,** 2251–2255.

Zewail, A. H., and Bernstein, R. B. (1988). *Chem. and Eng. News* **66,** 24–43.

7

THE TRANSITION STATE

John C. Polanyi
University of Toronto

The "transition state," broadly defined, encompasses all internuclear configurations of a reacting system intermediate between reagents and products; it embodies our understanding of the process of chemical change.

THE BARRIER

Though history has no beginning, we honour Svante Arrhenius as having initiated the study of the transition state (Arrhenius, 1889). Arrhenius expressed the sensitivity of reaction rate to temperature in the form

$$\text{Reaction rate} \propto \exp(-E_a/RT), \tag{1}$$

where E_a is the activation energy, a constant with the units of energy. He conceived of this as being the height of an energy barrier. Since it is often true that $E_a \gg RT$ (RT represents the mean molecular collision energy), reaction under thermal conditions must involve the tiny fraction of molecular collisions at T that have sufficient energy to cross the barrier of height E_a.

But how do the atoms move in that pivotal event? Gradually that question came to the fore.

BARRIER HEIGHT

It is one of the seeming paradoxes of scientific history that empirical knowledge of such a subtle quantity as the heights of energy barriers to chemical reactions preceded by some decades the measurement of the dissociation energies for the bonds in the stable molecules under attack, i.e., $D(B—C)$ in

$$A + B—C \rightarrow A—B + C. \tag{2}$$

When bond dissociation energies became known (Gaydon, 1968; Cottrell, 1958), the surprising result was obtained that (as a rule of thumb),

$$E_a \approx 0.1 \, D(B—C). \tag{3}$$

And so, without anybody really noticing it, the field of "reaction dynamics"—the study of molecular motions in the course of chemical reaction—was born, though the term itself only gained currency in the late 1950s.

I say that reaction dynamics was born, since Eq. (3) proves conclusively that chemical reaction, viewed at the molecular level, is a more subtle event than bond-breaking followed by bond-formation; that unimaginative scenario would require $E_a \sim D(B—C)$. If the process is not sequential, it must be concurrent, that is to say, the new bond must be forming at the same time as the old one is dissolving.

In a language that G. N. Lewis would have understood, the transition state must have a charge distribution that maximises bonding throughout. Crudely, but revealingly, one can write this as

$$A + B_0^x C \rightarrow [A \times B \circ C]^{\ddagger} \rightarrow A_0^x B + C. \tag{4}$$

The mystery of science extends beyond the question of who discovered what (which is less important) to the more profound question of how they discovered it. Often, all we can say is, as Einstein did of the general theory of relativity, that "the sense of the thing was evident." So it was with G. N. Lewis' electron-pair bond, and so too with an extension of that notion into the realm of reaction dynamics which we owe to Linus Pauling (Pauling, 1947) and to Harold Johnston (Johnston, 1966). (Pauling's classic *The Nature of the Chemical Bond* was, it should be recalled, dedicated to G. N. Lewis.)

What Pauling did that concerns us here was to recognise the value of identifying *fractional* chemical bonds by their "order," n. This quantity correlated with both the length and the strength of the bonds. Harold Johnston extended this concept into the domain that interests us, by postulating that the total amount of chemical binding was conserved during the course of a reactive encounter.

If the order of the bond being formed is written n_1 and that of the bond being broken as n_2, then, in the case that the atom being transferred is H, the postulate is that as n_1 changes by infinitesimals from zero to its final value of 1 corresponding to bound AH, n_2 changes from its initial value of 1 for HC to zero. Throughout the reaction $A + H—B \rightarrow A—H + B$,

$$n_1 + n_2 = 1. \tag{5}$$

By way of the relationship of n_1 and n_2 to the corresponding bond dissociation energies $D_1(n_1)$ and $D_2(n_2)$, this simple notion led to predicted barrier heights in accord with Eq. (3), and, most often, in agreement with experiment.

As the field of reaction dynamics developed it became evident that for an understanding of the preferred types of molecular motions, the

location of the barrier crest along the reaction pathway was as significant as its height. The Pauling–Johnston approach (called the "bond
energy bond order" method, or "BEBO") once again was found to embody the necessary information concerning barrier location. I shall refer to this later.

At the moment I merely want to underscore the fact that every genu-
ine insight into the nature of the physical world brings important dividends at a later date—dividends that could not have been anticipated
by the individual who invested an important part of her or his creative
life in that discovery.

That long-range return is the ultimate hallmark of discovery. And
its prevalence is the reason scientists resist the notion—so appealing
to bureaucrats—that fundamental science be selected according to its
foreseeable benefits, rather than, as it should be, according to its scientific worth.

Of course the objection to the scientific criterion is that it can only
be applied from within the priesthood of science—a priesthood which,
to its credit, does not claim infallibility. It is said that when Erwin
Schrödinger was sent de Broglie's Ph.D. thesis setting out the fundamental expression for the wave nature of matter, he kept it for two
weeks and returned it to the French mathematician Langevin with the
comment that it was rubbish (Jammer, 1966).

Whether this particular story is true or not, aspiring researchers
should be warned that the initial response to new ideas is likely to be
negative. The body of science protects itself from foreign tissue by a
process of rejection that you can see operating in the most eminent
scientific journals. Without this "immune-response" science would
have grotesque appendages growing all over the place. So, if a contem-
porary Schrödinger says that your work is rubbish, I beg you to remember that it could still be epoch-making, though it is probably
rubbish.

Schrödinger must have taken a keen second look at de Broglie's thesis, for two years later he proposed wave mechanics. That was 1926,
an historic year, the same year that a relatively unknown 25-year-old
arrived in Germany to work with Arnold Sommerfeld. The young post-
doc was Linus Pauling. This leads me to offer my final piece of advice
to aspiring scientists, which is to be born at the right time.

The occasion of Sommerfeld's 60th birthday in 1928 brings us firmly back to the subject matter of this paper. For it was that party 63 years ago that, unknown to most of the participants, marked the birth of the modern field of reaction dynamics. It was there that Fritz London in a brief paper presented the equation which is central to our understanding of the transition state (London, 1928).

For the reaction A + B—C → A—B + C, London stated that all the properties of the energy barrier—height, location and shape—are embodied in the quantum-mechanical expression for the potential energy $[V(r_{AB}, r_{BC}, r_{AC}) \equiv V(r_1, r_2, r_3)]$ given by

$$V(r_1, r_2, r_3) = Q_1 + Q_2 + Q_3$$
$$- \{\tfrac{1}{2}[(\alpha - \beta)^2 + (\beta - \gamma)^2 + (\gamma - \alpha)^2]\}^{1/2}, \quad (6)$$

where the Qs are "Coulombic" integrals (Q_1 is $Q_1(r_1)$, etc.) and the α, β, and γ terms are "exchange" integrals (α is $\alpha(r_1)$, β is $\beta(r_2)$, and γ is $\gamma(r_3)$). The Coulombic terms (speaking a little loosely) embody the energy of attraction between each proton and each electron, whereas the exchange terms (corresponding to the "resonance" integrals in Pauling's description of chemical bonds; see Pauling, 1940) are due to such quantum mechanical considerations as the delocalisation of the electrons and their indistinguishability.

The three Coulombic terms in the London equation cause the atoms to stick together, whereas the remainder of the terms, which can overpower the first three in configurations of close approach, make it necessary to do work to bring all three atoms together. It is these latter terms that express in the language of quantum mechanics the notion of saturation of chemical bonding, or "valence," which G. N. Lewis sought to explain in terms of electron-pair bonds. A quantitative description of the limited extent of chemical bonding is also a description of the energy barrier, central to reaction—and to this paper.

A remarkable aspect of this equation is that it decomposes the three-body potential, $V(r_1, r_2, r_3)$, into a series of integrals—the Coulombic and exchange integrals—each of which refers to one pair of atoms only. The equation was the answer to a prayer. In fact it was arrived at by an inspired process of groping which is the scientific equivalent of prayer. London published it without proof. Years later Coolidge and

James (1934) showed how one could derive it by subjecting the Schrödinger equation to life-threatening surgery. And yet the London equation is, as I said, of central importance to our understanding of the interaction potential in chemical reaction.

I mention the history of the equation not in order to demean it, but to remind you that the path of scientific discovery is more remarkable than we customarily acknowledge. What the London equation, for all its approximations, provides is an interpolation formula between the Morse function describing the binding in the reagent molecule B—C, and the Morse function for the product A—B. The interpolation formula has the special virtue that it is composed of the same building blocks that describe the quantum nature of binding in the reagents and products, for this, too, is a sum of Coulombic and exchange terms.

We are left, as with the Morse functions for the asymptotic species B—C and A—B, with the problem of finding suitable values for the Coulombic and exchange integrals empirically. In part we do this by using (experimental) spectroscopic data for stable B—C and A—C and then solving for the Coulombic and exchange terms in the same approximation as is implied by Eq. (6)—a procedure that we owe to Henry Eyring and Michael Polanyi (1931)—and in its more modern form to Sato (1955).

When, the year after London's equation was published, Eyring and my father introduced this notion, they thought that a consistent approximation to the Schrödinger equation applied along the entire journey from reagents to products would yield potential energies that, though they were not the true ones, would run parallel to the true energies. The calculated relief map of the terrain explored by the reacting species would then be the true map, but displaced upward by (so to speak) some thousands of feet relative to the actual sea level.

Since then the *ab initio* theorists (see, for example, Laidler and Polanyi, 1965, Chapter 1, p. 1, Table 1) have learnt, to their sorrow, that even in vastly better approximations than are represented by the London equation, a wave-function which is excellent asymptotically may or may not be excellent in the transition state region. This would be fine if one had any way of knowing whether the example being studied fell under the rubric of "may" or that of "may not." But one doesn't.

The Eyring–Polanyi–Sato expression is not, therefore, by itself, enough. To be confident about the geography of the potential surface, $V(r_1, r_2, r_3)$, one must send out an expedition to explore it.

Dudley Herschbach, in Chapter Eight of this volume, describes the scattering approach to the determination of features of the potential energy surface. The prime measurables are the angular distributions and speeds of the reaction products. This application of the methods of particle physics to chemistry represents a frontal attack on the problem of chemical change, under American generalship.

In Canada we take a more ruminative approach, warming our hands at the fire of life. We are encouraged in this by a government that appreciates the low cost of rumination. Our approach to reaction dynamics, starting in the 1950s, was to let the molecules formed in chemical reaction do the work by signalling to us their state of excitation. They did this through infrared emission (Cashion and Polanyi, 1958; Polanyi, 1987).

As a detector we had in the first instance something only slightly more sensitive than the palms of our hands, namely a thermocouple. Within a year, however, we were able to switch to semiconductor infrared detectors. They had been developed for the earliest heat-seeking missiles. For a time in the late 1950s they were unavailable to civilian users in the U.S., but, due to an oversight, could be obtained for export. There are advantages to a world of many nations.

Crossed uncollimated sprays of molecules entered a low-pressure $(10^{-6}$ torr) cryopumped vessel. Only a few collisions occurred at the crossing point. This was evidenced by the highly nonthermal state of vibrational and rotational excitation in the reaction products which gave rise to an infrared line-spectrum reflecting what we termed the "detailed rate constant," i.e. the distribution of newly born products over vibrational (V'), rotational (R'), and hence translational (T') states of excitation: $k(V', R', T')$.

We present these findings in the form of "triangle plots" (Polanyi, 1987) in which we plot contours of equal detailed rate constant in a vibrational, rotational, translational energy-space. The coordinates are V' along the ordinate, R' along the abscissa, and T' (which is the total energy in the products less $V' + R'$) increasing to a maximum at the lower left (where $V' + R' = 0$). The contours, which ride rough-shod over energy quantisation, are designed simply to assist in communicating the differing patterns of product energy distribution, $k(V', R', T')$, for different reactions.

The triangle plots have been further simplified in the accompanying figures by replacing the contours by coloured shading. What you see are hills (not on the potential-energy surface but) in "detailed rate-constant space." The space is limited by the energy release in going from $ABC^{\ddagger}$ at the crest of the energy-barrier to the product asymptote at $AB + C$ (plus a small contribution from the reagent energy—except in special experiments where this reagent component was deliberately enhanced by introducing excess energy of translation or vibration in the reagents, whereupon the triangular space available to the products was correspondingly increased). The peak of the hill—some two or-ders-of-magnitude higher than the plain—indicates the most probable vibrational–rotational–translational excitation in the reaction prod-uct. The breadth of the hill along the ordinate gives the vibrational distribution, that along the abscissa the rotational excitation, and that along a grid of diagonals the translational excitation.

This last is the same quantity that is measured in cross-beam ex-periments, but integrated here over all angles. In the infrared chemi-luminescence data, the product translational excitation is presented at a level of detail where it can be identified with individual states of vibration and rotation (Anlauf *et al.*, 1970). Molecular beam experi-ments are today approaching this level of specificity in the measure-ment of product energy-distribution, but have yet to attain it. So we remain dependent on this old data for the most detailed "fingerprints" identifying some exchange reactions, $A + B\text{—}C \rightarrow \overleftarrow{A}\text{—}B(v',J') + \overrightarrow{C}$ according to their product vibrational, rotational, and translational excitation.

BARRIER LOCATION

The potential energy function $V(r_1, r_2, r_3)$ that describes the tran-sition state is multivariant. In order to learn something about the po-tential function from the states into which the products are scattered, it is necessary to reduce the number of variables (see Polanyi, 1987, and references cited there). This was done by examining the properties of a range of different $V(r_1, r_2, r_3)$ and looking for features of overriding importance. Fortunately the classical mechanics of A + BC reactive scattering provides an excellent guide to the angular and energy distribution in the products.

M. G. Evans, H. Eyring, and M. Polanyi suggested over half a century ago that what today we recognise as "barrier-location" would be an important variable in determining product vibrational excitation. They used the classical equations of motion as their guide to this qualitative observation. In the absence of computers they could not solve the equations of motion, and in the absence of quantitative data regarding product energy distributions they had little incentive to do so.

With the serendipity that marks the history of science, the energy distributions that were needed and the high-speed computers required in order to solve the equations of motion arrived on the scene simultaneously in the 1960s. There is no magic about these happy coincidences (noted, in their generality, by Pangloss in Voltaire's *Candide*); it is simply that when they do not occur we do not record them.

The eight colour plates give some examples of $k(V', R', T')$ and their implications for the nature of the transition state. (I am indebted to my long-time associate, Mr. Peter Charters, for these coloured pictures.) The contrasting vibrational energy distributions for the reactions $H + Cl_2 \rightarrow HCl + Cl$ and $F + H_2 \rightarrow HF + H$ (Figs. 1 and 2) lead, through the solution of the equations of motion on trial $V(r_1, r_2, r_3)$, to the conclusion summarised picturesquely in Fig. 3. The energy-barrier is located "late" along the reaction pathway so that substantial energy release occurs as repulsion between the products. This "repulsive energy-release" gives a large amount of relative motion, i.e., translation, in the reaction products for the first reaction (for which the masses are $\underline{L} + \underline{H}\,\underline{H} \rightarrow \underline{L}\,\underline{H} + \underline{H}$, where $\underline{L}$ = light and $\underline{H}$ = heavy), whereas the same type of energy-release gives a high degree of internal motion, i.e., vibration, in the product molecule for the $\underline{H} + \underline{L}\,\underline{L} \rightarrow \underline{H}\,\underline{L} + \underline{L}$ mass combination.

To understand this one need only appreciate that the heavy red atom at the left in the top panel of Fig. 3 recoils slowly, carrying the blue atom (low mass) with it, whereas in the lower panel the blue atom at the left recoils with a high velocity, crashing into the red one to give high internal excitation before red and blue move slowly off to the left.

Figure 3 gives the lowest approximation to reality. In truth the (mean) transition state is somewhat bent, so that the repulsion applies a torque to the departing molecule, causing it to rotate. This is clearly evident in the experimental data for $F + HD \rightarrow HF + D$ as compared with $F + DH \rightarrow DF + H$ which shows on average twice as much rotational excitation in the HF product as compared with DF (Polanyi, 1987). Since the forces (i.e., the slopes of $V(r_1, r_2, r_3)$) are the same whether F approaches the D or the H end of HD, we are seeing here the effect of altering the sequence of masses while keeping the degree of bending in the transition state constant, as illustrated in Fig. 4. For reasonable $r_1^{\ddagger}$ and $r_2^{\ddagger}$ separations in the transition state, a bending of a little under $30°$ accounts nicely for the doubling of the mean rotational excitation in the product HF as compared with DF; repulsion as the system proceeds downhill from the crest of the energy barrier causes a higher velocity recoil in the H of HF (recoiling from the heavier D) than in the D of DF (recoiling from the lighter H), and hence more rotation in the former case.

To conclude this cursory survey, Fig. 5 shows a qualitatively different "triangle plot," which points an interesting moral for transition states. Figure 5 records $k(V', R', T')$ for the HCl product of the simple reaction $H + ClI \rightarrow HCl + I$. It is evident that, judged by its motions, there are two entirely different types of HCl being formed: HCl with low vibrational and rotational excitation (and hence high translational energy), and a larger amount of HCl with very high vibrational and rotational excitation (hence low translation). This suggests that in this example there are two pathways—effectively two different energy barriers—leading to the same chemical product.

We have termed this "microscopic branching" in order to distinguish it from the macroscopic branching (as in the case of alternative products HCl and DCl discussed above) for which the chemical nature of the products is different. Interestingly, microscopic and macroscopic branching are linked. The story of this linkage, stated briefly, is as follows.

In a reaction $A + B—C$, both B and C in the molecule under attack (since they bond with one another) can be expected to bond with the attacking atom A. We are dealing, therefore, with a quite general phenomenon. The presence of different products AB and AC is evidence of

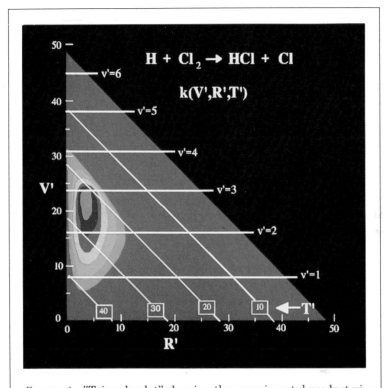

FIGURE 1. "Triangle plot" showing the experimental product vibrational, rotational and translational (i.e., V', R', and T') energy distributions for the reaction $H + Cl_2 \rightarrow HCl + Cl$, obtained
from infrared chemiluminescence. The most probable product
V', R', T' lies at the peak of the detailed rate constant hill $k(V'$,
R', $T')$, indicated by white. The value of $k(V'$, R', $T')$ in the
white region is ~100 times that at the dark blue sea-level. The
$k(V'$, R', $T')$ hill ascends through dark green (lowlands), pale
green, yellow, pink, and red, to the snowy region. The detailed
rate constant hill is located in a space which records increasing
vibration up the ordinate, increasing rotation along the abscissa
from left to right, and increasing translation (the balance of the
fixed *total* energy in the products) as diagonals increasing in
value toward the lower-left corner at which $V' = R' = 0$. This pic-
turesque rendering is based on a somewhat more quantitative
contour plot to be found in Polanyi (1987).

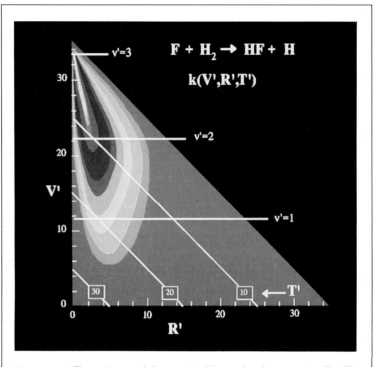

FIGURE 2. Experimental data as in Fig. 1, for the reaction $F + H_2$ → $HF + H$.

macroscopic branching. If, however, A moves rapidly or B and C separate slowly (due to large mass in B and C, or low repulsion) then a *given* product may be formed by two types of molecular encounter. Taking the case of AB as product, A may abstract B directly, or A may have an abortive reactive encounter with C which leads thereafter to "migration" of A from C to B, once again forming the chemical species AB but with a very different internal energy reflecting the very different sequence of forces that have operated on the emerging AB.

Figure 6 illustrates this for the experimental data given in Fig. 5. In Fig. 6 both a "direct" and an "indirect" (migratory) encounter are shown, leading in each case from H + ClI to the formation of the same products, HCl + I. The excitation in the product of direct reaction re-

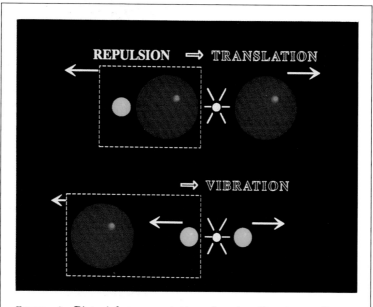

FIGURE 3. Pictorial representation of contrasting types of reaction dynamics for "repulsive" energy-release. The pictures are based on results obtained in 3D classical trajectory studies. The size of the spheres indicates their relative mass; the mass-combination above corresponds to $L + HH$ (e.g., $H + Cl_2$ for which the experimental findings were given in Fig. 1) and that below to $H + LL$ (e.g., $F + H_2$ for which the experimental findings are in Fig. 2). To an adequate first approximation, the energy release between the products can be regarded as occurring in a collinear configuration, as illustrated.

sembles qualitatively that for $H + ClCl \rightarrow HCl + Cl$ for which the experimental finding is shown in Fig. 1. The excitation from the indirect pathway resembles that (not shown here) for $Cl + HI \rightarrow HCl + I$ (Polanyi, 1987), a reaction with the $\underline{H} + \underline{L} H$ mass combination which, for the reasons evident in Figs. 3 and 4, gives high vibrational and (if bent) rotational excitation in the $\underline{H} \underline{L}$ reaction product. Since in a migratory encounter for $H + ICl$ the light atom spends some time in the region between the two heavy ones, the resemblance to $Cl + HI$ is not surprising.

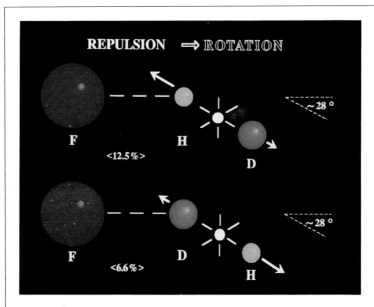

FIGURE 4. The message is similar to that in the lower panel of Fig. 3 (similar mass-combination), but a better approximation to the repulsive energy release is used in which the transition state is bent by 28°. The cases illustrated are for F + HD → FH + D (upper panel) and F + DH → FD + H (lower panel). Geometry and energy release are the same for these two isotopically related cases, but the latter is calculated to be only half as efficient in channeling translation into rotation than is the former. This is in agreement with the experimental findings (Polanyi, 1987).

The link between macroscopic and microscopic branching is evident from the relative heights of the two $k(V', R', T')$ mountains in Fig. 5. The reaction H + ClI favours approach from the I end, yielding more HI than HCl in the products of *macroscopic* branching. Accordingly, since abortive reaction with the I end gives the highly internally excited HCl, it is this *microscopic* branch that predominates in Fig. 5.

These plots of $k(V', R', T')$ for exoergic reaction have important implications for the reverse endoergic reaction, AB + C → A + BC.

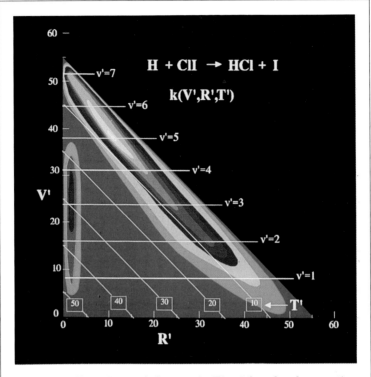

FIGURE 5. Experimental data as in Fig. 1 but for the reaction H + ClI → HCl + I; the product energy distribution is seen to be markedly bimodal.

With relatively minor corrections they yield $k_{endo}(V', R', T')$ for the uphill reverse reaction (V', R', T' are now the *reagent* energies) and tell us the optimal distribution of a fixed total energy over vibration, rotation, and translation in the reagents. This is shown for the endoergic reaction HF + H → F + H$_2$ in Fig. 7; the findings come from the application of detailed balancing to the data of Fig. 2 (Polanyi, 1987).

Inspection of trajectories across trial $V(r_1, r_2, r_3)$ for endoergic (uphill) reactions show that some translational energy is required to carry the system up the part of the energy barrier that lies along the coordinate of approach of AB to C in the endoergic reaction, and (most

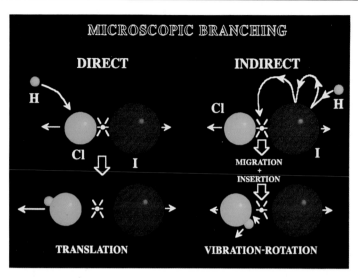

FIGURE 6. A simple pictorial rendering of two types of motions based on 3D classical trajectory modelling of a reaction of the type H + ClI in which both ends of the molecule under attack are highly reactive. The observed bimodal HCl vibrational-rotational energy distribution (Fig. 5) can be accounted for in terms of the "microscopic branching" pictured here; HCl with low internal energy (peak at the left of the triangle plot of Fig. 5) is due to *direct* reaction, and the HCl with high internal excitation (peak at the right of Fig. 5) is due to (the greater amount of) *indirect* reaction.

importantly) vibration is needed to surmount the remainder of the barrier that lies along the coordinate of *separation* of A from B. Both requirements are evident in Fig. 7, which is based on experiment.

It is easy to understand that motion of A away from B (corresponding to the coordinate of separation in endoergic reaction) can be more effectively brought about by placing vibration in the reagent AB ($\rightarrow\overleftarrow{A}$—$\overrightarrow{B}$) than by merely slamming AB into C, since AB vibration of the correct phase carries A away from B, as required.

Not only can we employ measurement of the relative efficiency of translation and vibration in inducing reaction as a probe of barrier location, we can also use this expedient as a practical means to select-

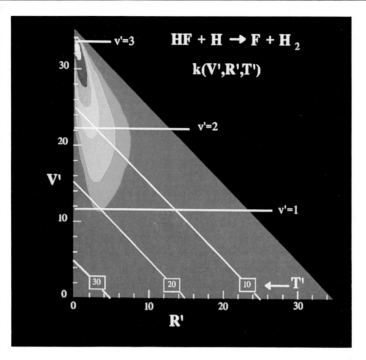

FIGURE 7. Triangle plot for the endothermic reaction $HF + H \rightarrow F + H_2$. The plot shows the dependence of the detailed rate constant on the *reagent* vibrational, rotational, and translational energies. The data were obtained by applying detailed balancing (time reversal) to Fig. 2. (Note that with the process running backwards, V', R', and T' refer to reagent energies.) The total energy in the reagents is fixed at the value that applied to the total energy in the products for exothermic reaction (Fig. 2).

ing reaction pathways. In the separation, let us say, of heavy from light hydrogen one can use a laser to excite vibrationally the deuterated or tritiated component of a naturally occurring mixture so that it reacts, leaving the unwanted hydrogenated isotope behind.

The drawback to this procedure is that infrared photons tend to be costly. This, in turn, may be offset by way of a second practical application of this work, namely the ease with which one can make vibrational lasers: as a result of the phenomenon of "partial population inversion," a partially cooled gas will lase on its P-branch transitions (Polanyi, 1987). Though this has nothing to do with reaction dynamics, it is a concept that emerged from thinking in regard to infrared chemiluminescence, which does have something to do with reaction dynamics.

Neither of these practical applications—vibrational lasers or vibrationally selective reactions—was anticipated. All that those involved in this field of research could offer their sponsors was the expectation that a significant improvement in understanding would be accompanied by a corresponding advance in the realm of application. All that they could *guarantee* was that if the pursuit of understanding was not allowed to proceed, with the consequence that discoveries were not made, significant new applications would not follow.

If this unexceptional proposition were to become widely accepted it would have profound implications for science policy in your country and mine.

BARRIER SHAPE

Under this somewhat ill-defined heading, brief reference will be made to some recent developments that are likely to result in a more detailed understanding of the form of the energy barrier or, equivalently, the nature of the "transition state." Foremost among these developments is a continual improvement in our ability to calculate $V(r_1, r_2, r_3)$ *ab initio* for simple cases from the Schrödinger equation (Siegbahn and Liu, 1978; Truhlar and Horowitz, 1978, 1979), and also to obtain the dynamics appropriate to this energy-surface from quantum scattering calculations (Zhang and Miller, 1989, and references therein). Nonetheless, for the great majority of reactions the need for empirical checks on the computations remains as great as ever.

A dozen years ago there was a vigorous debate as to the possibility of probing a reacting system while a molecular encounter was occurring, rather than directly after it was over. My laboratory's contribution to the debate was to point out that "the best evidence that something is possible is the fact that it has already been done" (Polanyi, 1979). In an important sense both the emission and absorption spectroscopy of systems engaged in collision had been the subject of study for the better part of a century, under the heading of "collisional line-broadening."

A collision perturbs the ground and the electronically excited states of the species that encounter one another differently. The electronic spectrum therefore, shifts to a new frequency, ν, with each change in separation between the collision partners, r. By measuring the distribution of absorption or emission intensities as a function of ν, $I(\nu)$, one is obtaining information directly concerning the relative times, t, that the colliding system spends at successive r, which we may call $P(t)$. This in turn embodies information regarding the relative slope of $V(r)$ (the force) operating at each r, and allows one to construct $V(r)$.

We can replace the collision that produces the classic collision-broadening with the half-collision that constitutes the separation of reaction products moving down the far side of the energy barrier (Polanyi, 1979). In this case the Fourier transform of the spectrum, $I(\nu)$, would yield the desired information, $P(t)$, for the species intermediate between reactants and products; we referred to this as "transition state spectroscopy" (TSS).

The example that we chose for an initial attempt at TSS was the reaction $F + Na_2 \rightarrow NaF + Na^*$ (where Na^* is in the upper state of the famous yellow D-line transition) which gives rise to intense luminescence at 589 nm (Polanyi, 1979; Arrowsmith et al., 1980, 1983). By filtering out this intense yellow line-emission in a succession of spectrographs, we found that we could record a feeble "wing emission" extending several hundred wavenumbers to either side of the D-line. The integrated intensity of these wings was $10^{-4} \times$ that of the D-line (whose lifetime is 10^{-8} s), implying a reactive collision of 10^{-12} s. This constituted a direct spectral measurement of the total lifetime of the transition state.

Notwithstanding its promise this approach, pursued in a number of laboratories, suffered from two evident weaknesses. In the first place, since $P(t)$ is the fundamental quantity of interest, it would be preferable to measure it directly. This has now been accomplished in elegant time-resolved experiments described in Chapter Nine by A. H. Zewail and R. B. Bernstein.

Current TSS experiments of a different sort are in the process of rectifying the second limitation of the early work. This limitation arises from the substantial averaging over reactive configurations which gives rise to a broad distribution over $P(t)$, washing out much of the structure in $I(\nu)$.

D. Neumark's laboratory, in a further remarkable development of TSS (see Chapter Nine, and Metz *et al.*, 1991), employs an emission spectroscopy (of electrons, not photons, but this is not essential to the approach) in which the transition state is accessed from a bound state ($ABC^- + h\nu \rightarrow ABC^\ddagger + e^-$). The favoured system for study has been the $\underline{H} + \underline{L}\,\underline{H}$ mass-combination (e.g., BrHBr) that parachutes onto the $\underline{H}\,\underline{L}\,\underline{H}^\ddagger$ barrier-crest in a configuration conducive to the to-and-fro motion of the light central atom between the heavy $\underline{H}$ outer atoms. The $\underline{H}\,\underline{L}\,\underline{H}^\ddagger$ is formed in various identifiable vibrational quantum states of this "chattering" motion, prior to falling apart to yield products $\underline{H}\,\underline{L} + \underline{H}$.

This "bound-to-free" TSS constitutes the most direct probe that we have at the present time of the shape of the potential surface perpendicular to the reaction coordinate. It will be fascinating to see this data employed in what has been termed and will now truly become an "absolute theory of reaction rates," ART. For the past half-century ART has provided the best statistical approach to the interpretation of measured rates of reaction, in which the rate is regarded as being governed by the number of ways in which the barrier can be crossed. But how real are the transition states this implies? We are about to discover.

This bound-to-free TSS, in common with the time-resolved TSS (see Chapter Nine), should prove to have broad applicability.

The time-resolved approach requires, clearly, that there be a defined initial configuration r_0 of the reagents corresponding to t_0, so that a time t following t_0 refers to the same separation r in the case of each of the multitude of molecular collisions under observation.

Until now the requirement for a defined initial configuration has been met by making use of gaseous clusters as the working material, a technique of high promise developed by Soep (Breckenridge *et al.*, 1987) and by Wittig (Shin *et al.*, 1991, and earlier references there).

The cluster contains a molecule DA that can be photolyzed to yield reagent A, and, in addition, it contains the second reagent BC. Since the cluster is bound by van der Waals forces in a preferred geometry DA--BC, the photorecoiling A encounters BC at a preferred angle to the B—C bond, at a preferred impact parameter (the distance by which A's trajectory misses the centre-of-mass of B—C), and following a trip across a preferred distance (namely the A--B separation in the cluster); together these parameters serve to define what was loosely described in the previous paragraph as r_0.

There is a second way in which these requirements for an aligned and ordered starting configuration, and hence a defined r_0, could be achieved, that has been the subject of study concurrently with the cluster method. This second approach to reagent ordering has not yet been applied to TSS, but seems very likely to be broadly applicable in that context.

This alternative approach falls under the heading of "surface-aligned photochemistry" (Polanyi and Rieley, 1991, and earlier references there). In the present instance it would involve co-adsorption of the molecule DA that can be photolyzed, along with a target molecule BC at some inert *crystalline surface*. For a number of cases A recoiling from adsorbed DA has been shown to react with the adsorbed BC, giving rise to A + BC → AB + C. In such a case A encounters BC at a preferred angle to the B—C bond, at a preferred impact parameter, and following a trip across a preferred distance—as illustrated in Fig. 8.

The principal advantage to this approach should lie in the greater degree of choice in regard to the relative positioning of DA with respect to BC—governed by the choice of substrate and coverage. A valuable bonus is the fact that the reacting species are aligned in the laboratory frame of reference so that the directions in which product is scattered will be particularly informative. The principal drawback, particularly if A is aimed downwards, is the proximity of the surface.

This can, however, be turned to advantage since it opens the way to an important extension of TSS into the realm of surface reaction. A photofragment recoiling from a molecule adsorbed on a preferred site will strike a particular type of surface atom at a preferred angle and impact parameter (in contrast to an impinging molecular beam which strikes all points on the surface with equal probability, and additionally has no characteristic separation r_0 of its atoms from the surface at time t_0).

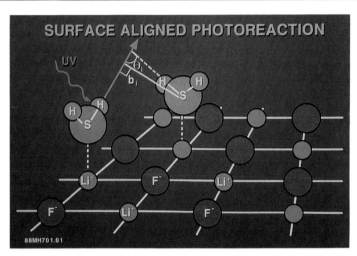

FIGURE 8. Schematic of "surface-aligned photoreaction." In the example shown, hydrogen sulfide is assumed to be adsorbed on a single crystal of lithium fluoride in register with the substrate atoms and with a preferred orientation relative to neighbouring H_2S. Photolysis leads to the ejection of an H along what was previously an S—H bond direction. This H approaches the neighbouring H_2S and after travelling a designated distance collides with an adjacent $H_2S(ad)$ with a restricted angle of approach (θ_i) and impact parameter (b_i).

In the example shown in Fig. 9, the impinging atom, having probed the interaction potential at a preferred site, scatters off the surface. Photolysis by a shorter wavelength, λ, gives rise to a correspondingly more energetic atom that probes the same localised site but, having penetrated more deeply into it, scatters at a different angle, θ. The variation of θ with λ maps out the potential-energy contours, V, at a particular atom on the surface. The extension to the case in which the attacking atom severs a chemical bond at the surface, presently under study in our laboratory is an obvious one.

It is an extension that is in tune with the times, since interest in reactions at surfaces is high today. It provides a goal for transition-state studies of the future, as well as involving the crystalline state that has been brought so vividly to life in the work of Linus Pauling.

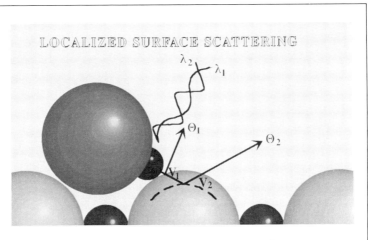

FIGURE 9. The spheres across the base of this figure represent Li$^+$ and F$^-$ (small and large hemispheres) at the surface of a LiF crystal. The adsorbed molecule, shown schematically, represents hydrogen bromide; the Br (large), being electronegative, is located over Li$^+$, whereas the electropositive H (small) is tilted down so as to form a hydrogen bond with an F$^-$ in the substrate. Incoming ultraviolet of longer wavelength, λ_1, dissociates the HBr and causes H to bounce off the outer perimeter of an F$^-$ (at a potential $\underline{V}_1$). The shorter wavelength ultraviolet, λ_2, gives rise to more energetic H that bounces off an inner potential contour of F$^-$ (at $\underline{V}_2$). It is evident that, through adsorbate photochemistry using tunable light, it should be possible to map out the potential-energy contours of a crystalline surface for scattering off a region on a selected atomic species. We designate this *localized surface scattering*.

As a great scientist and a global citizen, Linus Pauling has done more than live up to the highest standards of our profession; he has taught us how high those standards can be. He has not yet left science, but when he does so he will leave it enriched and future generations inspired. Meanwhile we shall look to him, as in the past, for leadership in science and also in the wider world that has been transformed by science. If we show the same degree of concern that he has shown, it is within our power to ensure that this vast transformation benefits all of mankind.

REFERENCES

Anlauf, K. G., Charters, P. E., Horne, D. S., Macdonald, R. G., Maylotte, D. H., Polanyi, J. C., Skrlac, W. J., Tardy, D. C., and Woodall, K. G. (1970). *J. Chem. Phys.* **53,** 4091.

Arrhenius, S. (1889). *Z. Physik. Chem.* **4,** 226.

Arrowsmith, P., Bartoszek, F. E., Bly, S. H. P., Carrington, T., Jr., Charters, P. E., and Polanyi, J. C. (1980). *J. Chem. Phys.* **73,** 5895.

Arrowsmith, P., Bly, S. H. P., Charters, P. E., and Polanyi, J. C. (1983). *J. Chem. Phys.* **79,** 283.

Breckenridge, W. H., Duval, M. C., Jouvet, C., and Soep, B. (1987). In *Structure and Dynamics of Weakly Bound Molecular Complexes* (A. Weber, ed.), p. 213. D. Reidel Pubs., Amsterdam.

Cashion, J. K., and Polanyi, J. C. (1958). *J. Chem. Phys.* **29,** 455.

Coolidge, A. S., and James, H. M. (1934). *J. Chem. Phys.* **2,** 811.

Cottrell, T. L. (1958). *The Strengths of Chemical Bonds.* Butterworths, London.

Eyring, H., and Polanyi, M. (1931). *Z. Phys. Chem.* B **12,** 279.

Gaydon, A. G. (1968). *Dissociation Energies.* Chapman and Hall, London.

Jammer, M. (1966). *Conceptual Development of Quantum Mechanics.* McGraw-Hill, New York. Quoted in Atkins, P. W. (1974). *Quanta.* Clarendon Press, Oxford.

Johnston, H. S. (1966). *Gas Phase Reaction Rate Theory.* Ronald Press Co., New York.

Laidler, K. J., and Polanyi, J. C. (1965). In *Progress in Reaction Kinetics,* Vol. 3 (G. Porter, ed.), Pergamon Press, Oxford.

London, F. (1928). *Probleme der Modernen Physik.* Sommerfeld Festschrift, p. 104.

Metz, R. B., Bradforth, S. E., and Neumark, D. M. (1991). *Advances in Chem. Phys.* **81,** Chapter 1, p. 1.

Pauling, L. (1940). *The Nature of the Chemical Bond.* Cornell University Press, Ithaca, New York.

Pauling, L. (1947). *J. Amer. Chem. Soc.* **69,** 542.

Polanyi, J. C. (1979). *Faraday Disc. Chem. Soc.* **67,** 129.

Polanyi, J. C. (1987). *Science* **236,** 680.

Polanyi, J. C., and Rieley, H. (1991). In *Dynamics of Gas–Surface Interactions* (C. T. Rettner and M. N. R. Ashfold, eds.), Chapter 8, p. 329. Roy. Soc. Chem., London.

Sato, S. (1955). *J. Chem. Phys.* **23,** 592, 2465.

Shin, S. K., Chen, Y., Nickolaisen, S., Sharpe, S. W., Beaudet, R. A., and Wittig, C. (1991). In *Advances in Photochemistry,* Vol. 16 (D. Volman, G. Hammond, and D. Neckers, eds.), p. 249. J. Wiley, Interscience, New York.

Siegbahn, P., and Liu, B. (1978). *J. Chem. Phys.* **68,** 2457.

Truhlar, D. G., and Horowitz, C. J. (1978). *J. Chem. Phys.* **68,** 2466.

Truhlar, D. G., and Horowitz, C. J. (1979). *J. Chem. Phys.* **71,** 1514(E).

Zhang, J. Z. H., and Miller, W. H. (1989). *J. Chem. Phys.* **91,** 1528.

Copyright © 1992 by Academic Press, Inc.
All rights of reproduction in any form reserved.
ISBN 0-12-779620-7

8

CHEMICAL REACTION DYNAMICS AND ELECTRONIC STRUCTURE

—•——◆——•—

Dudley R. Herschbach
Harvard University

INTRODUCTION: HISTORICAL PERSPECTIVE

As scientist, teacher, and citizen, Linus Pauling has created a splendid legacy, among the finest of this century. To take part in honoring his 90th birthday was a great joy. Like many physical chemists of my vintage, I am his scientific grandchild. My Ph.D. mentor was one of Linus' most distinguished scientific sons, E. Bright Wilson. For most of my academic career I have been privileged to be a colleague of Bright and two other extraordinary first-generation Pauling progeny: Bill Lipscomb and Martin Karplus. Our students have enlarged the Harvard branch of the Pauling genealogy by hundreds; however, the palm belongs to Martin and his wife Marci, whose son Mischa was born on Linus' 80th birthday!

My personal debt to Linus dates from 1950, when I was a freshman at Stanford. My advisor, Harold Johnston, who had received his Ph.D. in 1948 from Caltech under Don Yost, told me much about Linus and astutely introduced me to his excellent book, *General Chemistry.* It was captivating, a luscious antidote to the dull tome that served as our official text. I particularly liked Linus' blend of "real-life" descriptive chemistry with theory and his straightforward but magisterial style, augmented by the fine drawings by Roger Hayward. At the birthday symposium, about 40 years later, I had my copy autographed by Linus, telling him that I appreciate his book all the more now that I have taught freshman chemistry myself for nine seasons. As a senior at Stanford, I read the classic text by Pauling and Wilson, *Introduction to Quantum Mechanics,* for a course taught by Hal Johnston. That book, still in print, remains my favorite and most-used science text. About 15 years ago, I had my copy autographed by both Linus and Bright (tacitly admitting that I found quantum mechanics easier than freshman chemistry). As a graduate student, I studied another celebrated Pauling book, *The Nature of the Chemical Bond.* I was much impressed to see how Linus combined pragmatic empirical analysis with insightful theory to develop a comprehensive picture of structural chemistry. Also, I remember noting his generous remarks in dedicating the book to Gilbert Newton Lewis.

A few months before completing my Ph.D., I was startled to receive a letter from Linus. He invited me to visit Caltech and give two seminars, one dealing with my thesis work and one with what I intended to do in the future. Slyly, he said nothing to indicate that Caltech might be interested in me as a faculty candidate (nor had Bright given me any hint). I have vivid memories of that visit, almost exactly 33 years ago. Linus greeted me as if I were indeed his grandson, putting his arm around my shoulders as he ushered me into his office. It contained a profuse collection of molecular models, some with atoms nearly the size of basketballs and bonds like baseball bats. We discussed barriers to internal rotation of methyl groups. This was the topic of my first seminar, which described how Bright and his students determined such barriers from microwave spectra. The method exploited splittings produced by tunneling among the threefold potential energy minima for torsional reorientation of the methyl group about its axis. Linus had just written a paper pointing out that many aspects of the internal rotation barriers could be explained, at least qualitatively, by postulating substantial d-orbital hybridization in the C—H bonds.

The next day I described my fledgling plans and hopes for molecular beam studies of chemical reaction dynamics. My talk began with praise of the pioneering molecular beam experiments of Otto Stern; however, when I ignorantly referred to him as a physicist, Linus shouted out: "Otto Stern was a physical chemist!" Of course, it was so, and many times since I have enjoyed telling my research students how I learned that. At every subsequent meeting with Linus in later years, he has likewise given me insights into human as well as molecular chemistry.

This chapter deals with three topics: a sample of what molecular beam methods have enabled us to learn about reaction dynamics since my fledgling seminar at Caltech; a new experiment designed to orient reagent molecules; and aspects of an unorthodox electronic structure theory which takes the dimension of space as variable. In part, these topics illustrate current work and future hopes, just as stipulated by Linus in his letter, but they will also serve to celebrate three of his favorite themes: *electronegativity, hybridization,* and *resonance.*

First, however, I want to emphasize the historical context of reaction dynamics. Figure 1 indicates this for what might aptly be called the Pauling century of physical chemistry. In the beginning, more than a century earlier, thermodynamics emerged as the foundation for physical science. The thermochemical era reached its peak in 1923, with the publication of the classic text by Lewis and Randall. This was shortly before the discovery of quantum mechanics ushered in the new structural era. In turn, the grand pursuit of molecular structure and spectra reached a twin pinnacle in 1951 and 1953 with the discoveries of the alpha-helix by Pauling and the DNA double-helix by Watson and Crick. This was shortly before the early molecular beam and infrared chemiluminescence experiments appeared as harbingers of the dynamics era. It is striking that the prequantum models of Lewis and Irving Langmuir, still taught today, came well before the onset of the structural era. Likewise, the potential surface and transition-state concepts of Henry Eyring and Michael Polanyi anticipated the dynamics era.

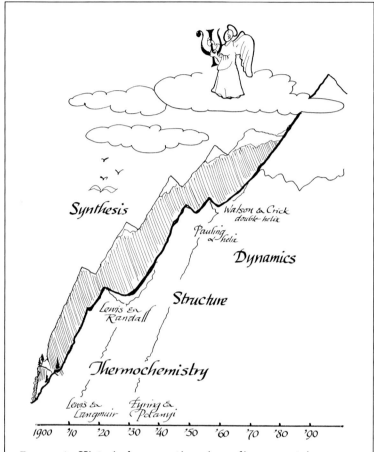

FIGURE 1. Historical perspective. Ascending mountains represent three eras, since *thermochemistry* was prerequisite for the
structure era and both underlie the *dynamics* era. All three con-
nect to a vast range representing *synthetic* chemistry, which
draws closest to dynamics as both disappear heavenwards into
clouds symbolizing the ultimate triumph of *ab initio* quantum
chemistry.

What a fabulous century of molecular science! Linus Pauling was
born soon after Planck discovered the quantum, began his scientific
work only a decade after Bohr proposed a magical model for the hydrogen atom and Laue devised x-ray diffraction, and received his Ph.D.

just as Heisenberg and Schrödinger were discovering quantum mechanics. Linus went off to Europe, mastered the new theory, and returned to make Caltech the mecca for structural chemistry. A fitting prologue to Pauling's crowning achievement, the deduction of protein structures from data and principles adduced from intimate study of simple molecules. Surely this and the kindred discovery of the DNA structure will be celebrated a millennium hence.

Not evident in Fig. 1, and in contrast to the scientific continuity, are cultural chasms between the eras. Linus once told me of the gulf his students encountered in the early 1930s when, as candidates for academic positions, they presented seminars describing molecular structures they had determined by electron diffraction. The physical chemistry faculty in the audiences had done their Ph.D.s in thermochemistry, and so imbibed a tradition which emphasized that it did not need to postulate the existence of molecules! Worse, interpretation of the diffraction rings then relied entirely on the so-called "visual method." The rings could only be seen because the human eye accentuates slight changes; a densitometer tracing showed just monotonically decreasing intensity. The molecular structures obtained by interpreting the spacing and heft of the rings in effect depended on an "optical illusion," incomprehensible and reprehensible to many professors of thermodynamics. More than 30 years later, my own students encountered a similar gulf when presenting seminars describing our early crude molecular beam reactive scattering studies. Almost all the incumbent faculty then had done their Ph.D.s in structure or spectroscopy. They were dubious about work that depended on drawing velocity vector diagrams to interpret bumps on scattering curves. Fortunately, such cultural chasms (although still wide) were soon bridged.

The elucidation of molecular structure and chemical bonding, so greatly advanced by Pauling, inspired hopes that something comparable could be achieved for reaction dynamics. I think this was vital in creating a sense of historical imperative among young physical chemists 30 years ago. An evangelical fervor like that so evident in Pauling sustained the mavericks who undertook to develop means to study individual molecular collisions. That likewise made them brash rather than abashed when confronting the many wise elders then skeptical or indifferent.

DYNAMICS OF ATOM TRANSFER REACTIONS:
ROLE OF ELECTRONEGATIVITY

As Otto Stern liked to emphasize, the great appeal of the molecular beam method is its conceptual "simplicity and directness." Applied to reactions, this involves forming the reagent molecules into two collimated beams, each so dilute that collisions within them are negligible. The two beams intersect in a vacuum and the direction and velocity of the product molecules emitted from the collision zone are measured. In the early days, the wise skeptics enjoyed pointing out an obvious paradox: "If collisions within each beam are negligible, when you cross the beams collisions will be still more negligible!" Fortunately, that paradox holds only as a first approximation, but intensity was a daunting problem. For the first several years, all the reactions studied successfully in crossed molecular beams belonged to a special family which yielded unusually high fluxes of product molecules. Even so, under feasible conditions the yield of product molecules at the detector corresponded to only *a monolayer a month,* generated from parent beams with intensities corresponding to a monolayer a minute.

The special family of reactions, so essential for the early beam experiments, all involved an alkali atom attacking a halogen-containing target molecule to form an alkali halide salt. Many of these reactions had been explored in the 1920s by Michael Polanyi. His method was simply to diffuse the reagents into opposite ends of a tube and estimate the overall reaction rate from the width of the salt deposit formed. He correctly inferred that the unusually large yields occur because the labile valence electron of the alkali atom readily switches to the halogen species, a process he called harpooning. Another vital advantage of these systems was the ease of detecting alkali species. This could be accomplished with high selectivity and efficiency by ionization on simple hot filaments, a detection scheme that goes back to Stern's laboratory.

Figure 2 shows an angle–velocity contour map for the potassium iodide salt formed in the reaction of potassium atoms with methyl iodide. This reaction was the subject of our first crude crossed-beam experiments, begun at Berkeley in the fall of 1960, which revealed that the product angular distribution was markedly anisotropic. Later the

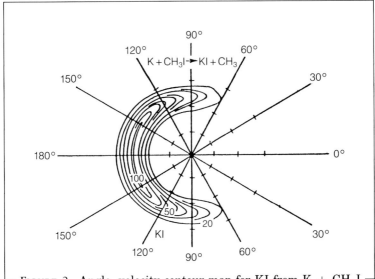

FIGURE 2. Angle–velocity contour map for KI from K + CH$_3$I reaction. Tick marks along radial lines indicate velocity intervals of 200 m/s.

reaction was studied repeatedly, serving as a prototype for testing many new experimental or theoretical methods devised for analyzing reactive collisions. In particular, lovely measurements of the product velocity distribution and other properties were carried out by Richard Bernstein and his students. The molecular beam studies also fostered extensive Monte Carlo "computer experiments," developed especially by Don Bunker, by Martin Karplus, and by John Polanyi, in which reaction properties are explored by tracing classical trajectories on a potential energy surface. This technique has proven invaluable for interpreting data and testing concepts of reaction dynamics. A review focused on the K + CH$_3$I reaction was given by Karplus to honor Pauling's 65th birthday.

The crescent shape of the KI product distribution in the contour map is typical for what is now termed a *rebound* reaction. An observer riding with the center of mass of the collision partners would see the product KI recoil into the backward hemisphere (and the CH$_3$ forwards) with respect to the incoming K beam. Markedly anisotropic scattering of this kind corresponds to a "hard collision" dominated by repulsive forces. The peak of the product velocity distribution, which varies little with the scattering angle, corresponds to release of about 50% of the reaction exoergicity (which is 25 kcal/mole, the difference in the I—CH_3 and KI bond strengths) into the translational recoil of KI relative to CH_3. The other 50% goes into vibrational and rotational excitation of the products. Predominant energy disposal in relative translation is typical for impulsive release of repulsion.

The electron-switch mechanism accounts nicely for the repulsion between the product molecules in this prototype rebound reaction. The harpooning electron enters a strongly antibonding orbital of methyl iodide which has a node between the carbon atom and the iodine atom. This is the same orbital populated in the excited electronic state that leads to photodissociation of methyl iodide. Therefore, we can estimate the repulsive energy release by examining the absorption spectrum of the parent molecule. The magnitude and even the distribution of the repulsion released in the alkali atom reaction indeed prove to be very similar to that in photodissociation. Dozens of other alkali reactions have been studied, with results amply reviewed elsewhere. For many the product yield and shape of the angle–velocity distributions differ greatly from the methyl iodide case, yet the variations and trends can be accounted for by examining the nature of the molecular orbital of the parent molecule which receives the harpooning electron. This chemical perspective was gratifying, but alkali reactions were considered an eccentric, unrepresentative family.

Happily, in 1968 reaction dynamics leapt "beyond the alkali age." This became possible with the advent of an extremely sensitive mass spectrometric detector, designed chiefly by Yuan Lee, who had arrived at Harvard about a year earlier to join our lab as a postdoctoral fellow. Yuan had already shown himself to be an extraordinary experimentalist in his Ph.D. work at Berkeley, and intended to try his hand at theory. But he was too shy to mention this to me when I asked him to undertake building the new apparatus. (I only learned of his hankering for theory 20 years later!) One of the first reactions we studied with the new machine was H + Cl_2, the reaction that John Polanyi had employed in developing his infrared chemiluminescence method. He had mapped out the energy disposal in vibration and rotation of the HCl product; our experiments provided the distributions of relative

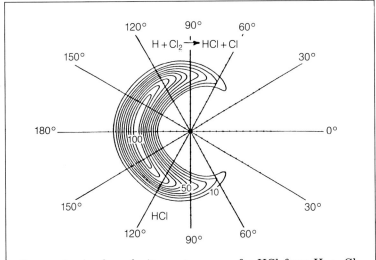

FIGURE 3. Angle–velocity contour map for HCl from H + Cl$_2$ reaction. Tick marks along radial lines indicate velocity intervals of 200 m/s.

translational energy and scattering angle. Figure 3 shows the product contour map we obtained. Except for a change of scale, the map for the H + Cl$_2$ reaction is almost congruent with that shown in Fig. 2 for the K + CH$_3$I reaction! It is uncanny that this pair of reactions, seminal in developing the chemiluminescence and molecular beam methods, proved to be so closely related; not only are both rebound reactions, they are even "kissing cousins." The 19th-century notation still used to write down chemical reactions gives no hint of such kinships.

The repulsive energy release for H + Cl$_2$ also proves to be very similar in magnitude and distribution to that found in photodissociation of the Cl$_2$ molecule. How this resemblance to photodissociation comes about is not obvious, since the very high ionization energy of the hydrogen atom prohibits electron transfer. We found a simple molecular orbital rationale using the "frontier orbital" concept of Fukui, and this led us to examine the role of electronegativity. Figure 4 compares the most rudimentary orbital descriptions for photodissociation and reaction. For simplicity we consider collinear approach of H to Cl$_2$ (the preferred reaction configuration, as shown by analysis of the HCl angular

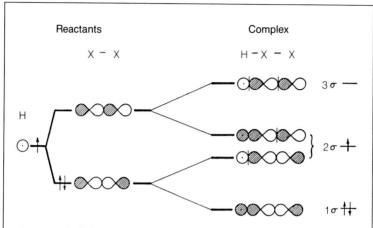

FIGURE 4. Schematic construction of the three σ molecular orbitals of collinear H–X–X. The *frontier orbital, 2σ,* may be regarded as a superposition of two components, one H–X antibonding and X–X bonding, the other vice versa.

distribution). This generates three σ orbitals for the H–Cl–Cl complex. The highest occupied orbital, denoted 2σ, is the frontier orbital. It has one node, resulting from the superposition of two components, one H–Cl antibonding and Cl–Cl bonding, the other vice versa. Simple semiempirical calculations indicate the latter component is dominant. Thus, the frontier node for the H atom reaction lies roughly midway between the chlorine atoms, just as in the lowest-lying unoccupied orbital of Cl_2, the orbital excited during photodissociation. This congruence in frontier nodes accounts for the strikingly similar repulsive energy release.

Other reactions of hydrogen atoms with halogen molecules provide instructive contrasts. As $Cl_2 \rightarrow Br_2 \rightarrow I_2$, the repulsive energy release becomes a smaller fraction of that in photodissociation, and the hydrogen halide angular distribution shifts from backwards to sideways with respect to the H atom direction. These aspects are illustrated in Fig. 5. The molecular orbital treatment related both trends to the decrease in halogen electronegativity. This enhances the p character of hybrid orbitals involving the central atom and thereby favors a bent configuration. The frontier node also shifts from about midway between the halogen atoms in H–Cl–Cl to close to the central atom in H–I–I. Many analogous stable molecules are known which have one more or one less valence electron than these H–X–X systems. The molecules with one more electron are linear or nearly so; those having one less electron are strongly bent, with bond angles of 90° to 110°. The shift of the frontier node might be expected to make H–I–I resemble a case with one less electron. Thus, it is plausible that decreasing the halogen electronegativity fosters bent reaction geometry and reduces the repulsive energy release.

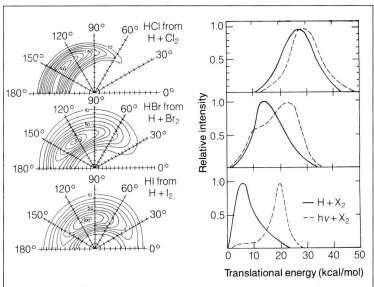

FIGURE 5. Comparison of angle–velocity contour maps for reactions of H atoms with halogen molecules. Since maps must be symmetric with respect to the horizontal axis, only upper halves are shown. Tick marks along radial lines indicate velocity intervals of 200 m/s for Cl_2 case, 100 m/s for Br_2 and I_2. Panels at right compare distributions of product relative translational energy (solid curves) with continuous absorption spectra of halogen molecules (dashed curves). Abscissa scales for spectra are shifted to place origin at the dissociation asymptote, and thus show directly the repulsive energy release in photodissociation.

To study further the role of electronegativity, we looked at several reactions of the ICl molecule. For instance, the H + ICl reaction is much more exoergic to form H—Cl (bond strength 102 kcal/mol) than to form H—I (70 kcal/mol). Hence, on an energetic or statistical basis, reaction at the "Cl-end" would be more favorable than at the "I-end." By polling chemists at seminars and meetings, I found that almost all expected the reaction to go predominantly via the Cl-end; that opinion persisted even years after we had published experimental results demonstrating the contrary. The molecular orbitals, shown in Fig. 6, suggest the H atom should prefer to attack the I-end. As a consequence of the electronegativity difference, in ICl both the highest occupied orbital (π^*) and the lowest unoccupied orbital (σ^*) are predominantly I atom orbitals, so reaction should at least be initiated at the I-end. This orbital asymmetry accounts for several striking features of the spectra of interhalogen molecules, as Mulliken pointed out long ago. After our beam study of H + ICl found the yield of HI is at least comparable to HCl and probably substantially higher, I phoned John Polanyi in an exuberant mood to urge him to look at the infrared chemiluminescence from this reaction. He was initially a bit reluctant, saying his troops were weary of halogen reactions, and noting that he would not be able to see emission from HI because its dipole derivative is exceptionally small. However, he did study the chemiluminescence from HCl and was rewarded with lovely results. The energy distribution in vibration and rotation has a very unusual bimodal form which revealed that some of the HCl is formed by direct attack at the Cl-end, but four times as much comes from indirect reaction as the H atom pinwheels about after initial attack at the I-end.

Analogous steric preferences for many other atom transfer reactions can likewise be attributed to orbital asymmetry arising from differences in electronegativity. As seen in Fig. 6, no matter whether the attacking reagent wants to remove or to insert an electron (or some lesser partial charge), the target orbital involved favors the I-end of the ICl molecule. Indeed, beam studies found that as with H atoms, reactions of Br, O, and CH_3 with ICl all form predominantly iodides. This is remarkable, since these various reactions differ in many other ways. From analysis of triatomic molecular orbitals, there emerges a

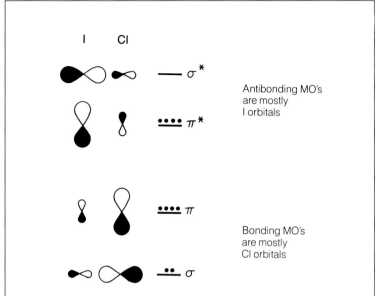

FIGURE 6. Molecular orbitals for ICl, showing asymmetry resulting because the Cl atom is more electronegative than the I atom.
Bonding orbitals σ and π are composed predominantly of Cl
atomic orbitals, conjugate antibonding orbitals σ^* and π^* predominantly of I atomic orbitals.

general criterion which we call "the electronegativity ordering rule."
This predicts that H–X–Y will be a more favorable conformation than
H–Y–X, if the X atom is less electronegative than Y (where these are
any atoms with p electrons). Likewise, in X–Y–Z systems, the preferred conformation has the least electronegative atom in the middle.
This rule holds for almost all known stable triatomic molecules, linear
or bent, and appears to apply also to preferred reactive conformations.

An especially striking example of the electronegativity ordering rule
was found in reactions of O atoms with halogens. Whenever the O
atom is more electronegative than one or both halogen atoms in the
target molecule (i.e., Cl, Br, or I), the reaction goes readily, with no or
very low activation energy. For mixed halogens XY with X less electronegative than Y (e.g., ICl or ClF), only the OX product appears and

not OY, despite the much larger exoergicity for the latter path. This is nice evidence that the O atom attacks end-on rather than inserting between the halogen atoms. However, since oxygen is less electronegative than fluorine, the O + F_2 reaction should prefer the F–O–F geometry rather than O–F–F. This implies a relatively high activation energy, associated with the switch to an insertion mechanism, despite the large reaction exoergicity and the notorious chemical personalities of the reactants. In my informal polling of chemists, I found nobody who believed this unorthodox prediction; all thought that O + F_2 would be at least as facile as the O + Cl_2 reaction. Subsequent experiments indeed confirmed that O + F_2 is inhibited by a large activation energy (~12 kcal/mol); the rate at room temperature is about seven orders of magnitude slower than the O + Cl_2 reaction.

When the electronegativity difference between the attacking and target atoms becomes so large that electron transfer occurs, as in harpooning reactions of alkali atoms, other instructive considerations enter. For instance, in the K + ICl reaction, the major product is KCl rather than KI. The key feature here is that the electron switch occurs at large distance (~7 Å), so the actual reaction involves a K^+ cation incident on an $(ICl)^-$ anion. Since the harpooning electron should enter the σ^* orbital of Fig. 6, in the intermediate molecular anion this electron initially resides mainly on the I atom, but usually will shift to the more electronegative Cl atom as the $(ICl)^-$ anion dissociates in the field of the incoming K^+ cation. Auxiliary evidence supports this interpretation. For a variety of charge-transfer complexes with ICl, structure studies show that the donor group is adjacent to the I atom and the I—Cl bond distance is expanded, as expected when charge is deposited in the σ^* orbital. The nonreactive scattering of alkali atoms from ICl shows distinctive features that indicate reaction is inhibited or precluded for an appreciable fraction of collisions at small impact parameters. These collisions may involve configurations in which the I (or I^-) atom blocks the access of K (or K^+) to the Cl atom.

Electronegativity differences also have a prominent role in reactions involving transfer of two atoms. In accord with the celebrated Woodward–Hoffmann rules, the venerable textbook reaction H_2 + $I_2 \rightarrow 2HI$ does not occur as an elementary process; instead it involves dissociation or near-dissociation of I_2 followed by I + H_2 + I. Many other four-center bimolecular reactions are likewise forbidden as concerted processes. However, when one or two of the atoms involved differ greatly in electronegativity, four-center reactions can become quite facile. This we demonstrated for several cases involving ionic bonds. A favorite example is $Cs^+Br^- + ICl \rightarrow Cs^+Cl^- + IBr$. The reaction very likely involves formation of an alkali trihalide salt, $Cs^+(BrICl)^-$, and charge migration within the trihalide ion. Although apparently unknown in the gas phase, trihalides have been much studied in solution and in the solid state. As expected from molecular orbital theory, the trihalide anions are linear or nearly linear, the middle atom is always the least electronegative (I, in this case), and it acquires a fractional positive charge, whereas the end atoms share the negative charge. Clear evidence for this structure appears in the reactive scattering. There is no observable yield of $Cs^+I^- + BrCl$, even at collision energies more than 20 kcal/mol above the energetic threshold for this channel. Furthermore, as seen in Fig. 7, the IBr angle–velocity contour map has a very unusual skewed shape. The left-hand product peak has distinctly higher intensity and velocity than the right-hand peak. This shows that collisions from which IBr and CsCl rebound backwards with respect to the incident ICl and CsBr, respectively, are more probable and

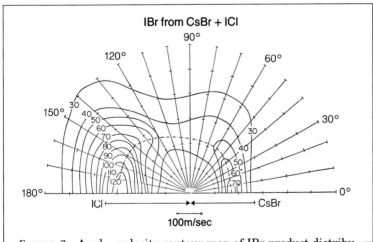

FIGURE 7. Angle–velocity contour map of IBr product distribution from CsBr + ICl reaction.

involve more repulsive energy release than collisions from which the IBr and CsCl emerge in the same direction as the incident ICl and CsBr, respectively. These properties are consistent with the expectation that in Cs^+Br^- + ICl reactive configurations, Br^- tends to be collinear with ICl while Cs^+ is likely to collide with the central I atom. The partial positive charge acquired by I as the trihalide forms then repels Cs^+, which picks up the emerging Cl^- and departs in the direction opposite to the incident salt.

With the grand saga of molecular structure in view, from the outset my own chief goal in pursuing "single-collision chemistry" has been to study how reaction dynamics is governed by electronic structure. The brief sampling given here indicates how I have tried to emulate Linus Pauling in seeking qualitative rules, rationalizations, and semiempirical correlations. With this emphasis on electronic structure, as exemplified in his work, insights derived from reaction dynamics can be applied to chemical phenomena far beyond the realm of collision chambers and computers.

ORIENTING PENDULAR MOLECULES:
HYBRIDIZED ROTATIONAL STATES

No matter how much care is devoted to defining or analyzing directions, velocities, and internal states of the collision partners, the dynamical resolution of reactive scattering experiments is limited by two random aspects of the initial conditions. These are the "dartboard" distribution of impact parameters in the collision, and the random orientation of the target molecule. Both are what I call "Garden of Eden problems"; that is, they keep us from acquiring tantalizing information or "forbidden fruit." Comparing trajectory calculations performed with and without averaging over these random distributions shows that the reaction dynamics could be characterized much more incisively if the averages could be undone or avoided. In fact, this is now feasible.

Figure 8 indicates how the averaging over impact parameter can be undone. The impact parameter **b** denotes the closest distance of ap-

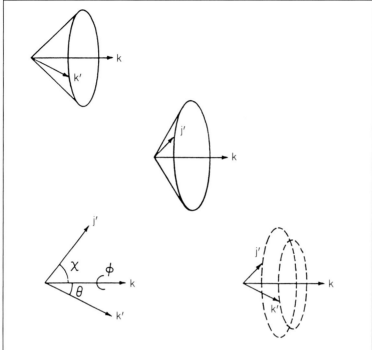

FIGURE 8. Three-vector correlation among initial and final relative velocity vectors, denoted by **k** and **k′**, and product rotational
angular momentum vector **j′**. Upper pair of diagrams indicate
the azimuthal symmetry about **k** of the **k′** and **j′** vectors inherent when these are observed separately, as in the two-
vector correlations (**k**, **k′**) and (**k**, **j′**). Lower pair of diagrams
indicates how the three-vector correlation (**k**, **k′**, **j′**) can give information about the dihedral angle φ, in effect undoing the azimuthal averaging about the initial relative velocity.

proach of the collision partners, in a hypothetical encounter with the
forces switched off. Both the magnitude of **b** and its azimuthal orientation about the initial relative velocity vector **k** are random, but here
we are concerned just about the azimuthal orientation. The magnitude
of **b** is in effect observable, since it enters directly into many collision
properties. For instance, a head-on collision has $b = 0$ and the parti-

cles scatter at wide angles; a grazing collision has b large compared with the range of the forces, so the particles scatter at small angles. However, the random azimuthal distribution of impact parameters makes the final relative velocity vector $\mathbf{k}'$ azimuthally symmetric about the initial velocity vector $\mathbf{k}$, although the forces acting in the collision do not usually have such symmetry (unless both partners are spherical). In the same way, darts thrown at a board exhibit azimuthal symmetry even when the thrower has astigmatism. Likewise, the distribution of any other vector property of the collision such as the final rotational angular momentum $\mathbf{j}'$ of a product molecule must be azimuthally symmetric about $\mathbf{k}$. The redeeming strategy is to measure two product vectors such as $\mathbf{k}'$ and $\mathbf{j}'$ simultaneously and thereby to determine the *triple-vector correlation* among $\mathbf{k}$, $\mathbf{k}'$, and $\mathbf{j}'$.

When a subset is selected of $\mathbf{k}'$ vectors with particular $\mathbf{j}'$ (or vice versa), this subset in general will not have azimuthal symmetry about $\mathbf{k}$. Accordingly, the dihedral angle ϕ between the $\mathbf{k}$, $\mathbf{k}'$ and $\mathbf{k}$, $\mathbf{j}'$ planes need not be uniformly distributed. Evaluating the distribution of ϕ recoups information about the reaction dynamics that otherwise would be lost by the azimuthal average over impact parameters.

As a chemical analogy, consider a helium atom in its ground state. The distribution of each electron, viewed separately, is spherically symmetric. But if the two electrons are observed simultaneously, their positions are strongly correlated as a consequence of their repulsive interaction. If not known beforehand, the presence of this repulsion would be revealed by the simultaneous observation, although not by viewing the electrons singly.

Although it is convenient to use the product rotational angular momentum $\mathbf{j}'$ as the "extra" quantity in the triple-vector correlation, this poses an instructive question. According to quantum mechanics, only the magnitude and one projection of an angular momentum vector can be specified. Since we envision a measurement that specifies both the magnitude of $\mathbf{j}'$ and the polar angle χ between $\mathbf{j}'$ and $\mathbf{k}$, the azimuthal angle about $\mathbf{k}$ should be unobservable. Is not the dihedral angle ϕ therefore unobservable after all? That is so, for any particular measurement. But this can be circumvented by measuring the angular momentum distribution using several different choices for the axis of quantization. The data can then be combined to obtain moments of the

ϕ distribution as well as those of χ and the scattering angle θ between the **k** and **k'** velocity vectors. The classical version of the three-vector correlation hence can be resurrected, including its ϕ dependence. In this sense, quantum mechanics in effect allows a dihedral angle such as ϕ to be observed even when the uncertainty principle does not permit this in any particular measurement.

This undoing of impact parameter averaging in molecular collisions has some heuristic resemblance to the celebrated "phase problem" encountered in x-ray crystallography. Both are resolved by introducing additional observables that provide new reference points, and by combining data from variant experiments.

The random orientation of target molecules is a more obvious limitation on the stereodynamical resolution of collision experiments but has remained a recalcitrant problem. Several techniques have been developed for aligning the rotational angular momentum **j** of a reagent molecule, but these are quite limited in chemical scope. The only means for orienting a molecule itself (rather than just its axis of rotation) has been electric field focusing, but this is applicable only to certain rotational states of symmetric top molecules. The method, pioneered 25 years ago by Bernstein and by Brooks, has enabled elegant studies to probe the orientation dependence of symmetric tops colliding with reactive atoms, photons, or surfaces. The focusing field technique depends on the fact that symmetric tops in certain states precess rather than tumble. As a result, the dipole moment does not average out but rather has a constant projection on the field direction. State-selection by the focusing field thus suffices to pick out molecules with substantial orientation of the figure axis. Orientation of polar molecules other than symmetric tops by an electric field has been considered to be quite impractical. The rotational tumbling of diatomic, linear, or asymmetric top molecules averages out the dipole moment in first order and hence greatly weakens interaction with an electric field. Many papers contain a ritual statement lamenting this situation: e.g., "Unfortunately, for diatomic molecules one requires extremely high electric fields to achieve even slight net orientation."

This long-held conventional wisdom has now been happily refuted. In considering orientation of nonsymmetric tops, the key parameter is $f = \mu\varepsilon/E_{rot}$, the ratio of the interaction energy of the dipole moment μ

with the electric field ε to the rotational energy of the molecule, E_{rot}. Typically, for a gas at room temperature, this ratio $f \approx 0.01$ to 0.001, whereas a ratio $f \geq 1$ is required in order to produce substantial orientation of the dipole. In a molecular beam formed by supersonic expansion, however, rotational relaxation can drop the rotational temperature down to 1 K or less. We can thereby obtain favorable ratios $f \geq 1$ for many molecules without need for inordinately large field strengths. For large f, the molecule can no longer rotate through $360°$ like a pinwheel; it becomes bound to the field direction and in this oriented state can only liberate over a limited angular range like a pendulum.

It is curious that this simple strategy was so long overlooked. The conversion of pinwheel rotation to pendular libration by lowering temperature was discussed more than 60 years ago by Linus Pauling in a paper dealing with rotation of molecules in crystals. Also, the quantum mechanics of a molecular pendulum is akin to that for the internal rotation of a methyl group, the subject of my first seminar 33 years ago at CalTech. The drastic cooling attainable in supersonic molecular beams has been well understood and exploited for 20 years. Yet somehow the virtue of combining these venerable ingredients was not recognized until just about a year ago.

For a rigid linear dipole μ interacting with a uniform electric field ε, the Schrödinger equation is

$$[B\mathbf{J}^2 - \mu\varepsilon \cos \theta]\psi = E\psi, \tag{1}$$

with B the rotational constant of the molecule (inversely proportional to the moment of inertia), $\mathbf{J}^2$ the squared angular momentum operator, E the rotational energy, and θ the angle between the molecular axis and the electric field direction. The dimensionless ratio $\omega = \mu\varepsilon/B$ specifies the strength of the interaction. The lower the value of J and the larger the value of ω, the better is the prospect for orienting the dipole by binding it to the field direction and the narrower is the librational amplitude of the molecular pendulum.

Figure 9 provides a nomogram which shows how the rotational quantum number J at the peak of the Boltzmann distribution varies with the rotational constant B of the molecule and the rotational temperature. This is still applicable in the presence of a strong ε-field,

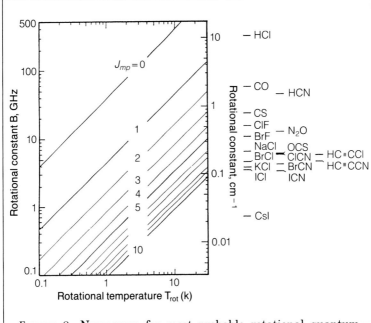

FIGURE 9. Nomogram for most probable rotational quantum number J_{mp} in Boltzmann distribution for a linear rotor as function of rotational temperature T_{rot}(K) and rotational constant B (GHz units on left scale, cm^{-1} on right). Values of B for typical polar molecules indicated at right.

since the initial population of field-free J-states is adiabatically transformed into the pendulum/pinwheel states. Values of B are indicated for typical molecules. We see that if we cool molecules such as ICl or ICN to 1 K in a supersonic beam, the most populated state will have $J \sim 1$; about 60% of the beam will have $J \leq 2$ and 95% will have $J \leq 4$.

Figure 10 shows another nomogram which displays the accessible range of the ω-parameter for a variety of linear polar molecules. The value of ω required to make all states with J less than or equal to a designated value $J_<$ become pendular states with librational amplitudes $|\theta| < 90°$ may be found from tabulated eigenvalues for the spherical pendulum problem. Values of $\omega \geq 15, 40, 67, 100$, respectively, suffice to bind all rotor states with $J_< = 1, 2, 3, 4$, respectively. Field

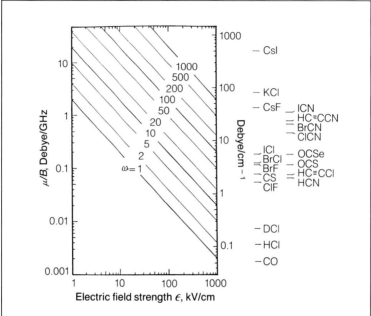

FIGURE 10. Nomogram for dipole interaction parameter $\omega = \mu\varepsilon/B$ as function of electric field strength, ε (kV/cm), and ratio of dipole moment to rotational constant, μ/B (Debye/GHz units on left scale, Debye/cm^{-1} on right). Values of μ/B for typical molecules indicated at right.

strengths up to $\varepsilon \sim 100$ kV/cm or a few times higher appear feasible. Thus, sizable values of ω, typically well above 10, can be attained for molecules with μ/B near or above that for OCS or ICl. Attaining large ω is futile if B is very small, unless the rotational temperature can be made very low. For instance, CsI requires a high source temperature, so T_{rot} cannot be made very low; many rotational states will be populated, only a small fraction of which can be oriented. On the other hand, when B is very large, as for HCl, essentially only the ground-state $J = 0$ will be populated; the whole beam can then be oriented, but only weakly because the feasible value of ω is small. However, there are many quite favorable cases. For instance, $\omega \sim 100$ appears accessible for the ICN molecule.

Typical energy level patterns and angular probability distributions are depicted in Fig. 11. For simplicity, here we consider a planar pendulum/pinwheel problem, obtained by replacing $\mathbf{J}^2$ by its two-dimensional version, $-d^2/d\theta^2$. This serves to illustrate the major features and allows us to obtain the energy levels and wavefunctions from

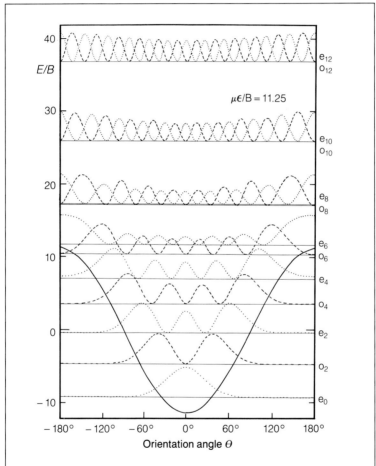

FIGURE 11. Energy levels (in units of the rotational constant B) and probability distributions, $|\psi(\theta)|^2$, for linear polar molecule treated as a rigid planar pendulum/pinwheel. Derived from Mathieu function tables (with tabulation parameter $s \equiv 8\omega = 90$).

tabulated solutions of the Mathieu equation. The same dodge was adopted by Linus in his treatment of molecular rotation in crystals. The cosine interaction potential of Eq. (1) provides an attractive well in the region $|\theta| < 90°$, flanked by repulsive barriers at wider angles. Thus the bound states created by the electric field are of two kinds: Those at negative energy (below the $J = 0$ field-free level) reside in the attractive potential well; those at positive energy are trapped within the repulsive barriers. Both kinds are pendulum states, but the librational amplitude is below $\pm 90°$ only for the negative energy states. The depth of the well and the height of the barriers increase in proportion to the field strength, so the number of pendulum states held by the potential increases with the ω-parameter.

For the example shown, there are six pendulum states: three bound with negative energies (labeled e_0, o_2, e_4 in customary Mathieu function notation) and three (o_4, e_4, o_6) with positive energies. All others are unbound pinwheeling states: one state (e_6) located just above the top of the $\cos \theta$ barrier, and higher states (o_8, e_8, . . .) occurring as nearly degenerate pairs that increasingly resemble the field-free planar rotor states. However, the angular motion is not uniform in these states above the barrier. The tumbling dipole speeds up as it swings through the attractive range ($|\theta| < 90°$) and slows down as it traverses the repulsive barriers ($|\theta| > 90°$). On average the pinwheeling dipole then points the "wrong" way; for these states $\langle \cos \theta \rangle$ is negative, corresponding to net repulsion. This perverse effect, pointed out long ago by Otto Stern, becomes quite pronounced for states (such as e_6) only slightly above the barrier. Even the bound pendulum states with positive energies favor the repulsive region. Only in the bound states with negative energies is the librating dipole limited to the attractive range where $\langle \cos \theta \rangle$ is positive. The net distribution of the orientation angle θ is given by the sum of the probability distributions for the various states, each weighted with its rotational Boltzmann factor. Thus, to attain substantial net orientation, with $\langle \cos \theta \rangle$ appreciably positive, it is essential that the rotational temperature be low enough to allow the pendulum states with negative energies to outweigh the higher states.

Figure 12 shows polar plots of the Mathieu eigenfunctions for zero field and for two values of the ω-parameter well within the range accessible for many molecules. For the field-free planar rotor, the wavefunctions for angular momentum J are simply $\cos J\theta$ and $\sin J\theta$. The

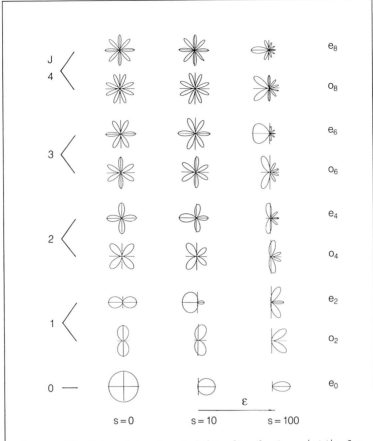

FIGURE 12. Polar plots of probability distributions, $|\psi(\theta)|^2$, for wavefunctions of a linear polar molecule treated as a planar pendulum/pinwheel. States for $s = 0$ pertain to field-free planar rotor with angular momentum $J = 0, 1, 2, \ldots$; these are simply circular harmonics, $\cos J\theta$ and $\sin J\theta$. States for $s \equiv 8\omega = 10$ and $s = 100$ indicate hybridization of the J-states induced by increasing electric field strength.

oriented dipole states created by the interaction potential are hybrids of the field-free states; thus they comprise Fourier series,

$$\psi(\theta; \omega) = \sum_m A_m \cos m\theta \quad \text{or} \quad \sum_m B_m \sin m\theta, \qquad (2)$$

where the sums (over $m = 0, 1, 2, \ldots$) pertain to the even and odd states, respectively. The hybridization coefficients A_m and B_m for a given pendulum or pinwheel state depend solely on the ω-parameter. In accord with the uncertainty principle, the narrower the amplitude of θ in any state, the broader the range of the field-free J-states that contribute significantly to the hybrid wavefunction. Of course this is just a planar analogue of the atomic orbital hybridization introduced by Linus long ago. Indeed, as already noted, my first conversation with Linus was about his 1958 paper attributing the origin of hindered internal rotation of methyl groups to the contribution of d-orbitals in the C—H bond hybrids. His argument invoked the substantial sharpening of the directionality of the C—H bonds produced by mixing in d-character.

Figure 13 illustrates for the case of the $\psi(e_2)$ state how the composition of the hybrid wavefunction changes as ω is increased. For the field-free rotor, this state has $J = 1$ and $\psi(e_2)$ proportional to $\cos\theta$, analogous to an atomic $p\sigma$-orbital. However, as ω rises, this $J = 1$ component declines steadily, whereas contributions from the isotropic $J = 0$ state (parent of e_0) and four-lobed $J = 2$ state (parent of e_4) grow in, reach maxima, and decline in turn as still higher J-states (e_6, e_8, . . .) enter the hybrid. For $s = 8\omega$ less than ~10, the $\psi(e_2)$ hybrid corresponds roughly to an out-of-phase overlap of the isotropic and $\cos\theta$ components, complementary to the in-phase combination forming the ground-state $\psi(e_0)$ hybrid. These two hybrids are analogous to antibonding σ^* and bonding σ molecular orbitals and point in opposite directions. At higher fields, the $\cos 2\theta$ component and others enter strongly and the parent $\cos\theta$ component practically disappears. This converts $\psi(e_2)$ into a three-lobed hybrid oriented with the field, analogous to a bonding σ orbital. The reversal of orientation arises in the same way as the "boomerang" Stark effect seen in asymmetric top molecules when the perturbation mixes three or more rotational states.

Figure 14 shows how the rotational spectrum of a linear dipole changes drastically when subject to a strong field. The spectrum in the absence of the field is just the familiar progression of equally spaced transitions, with successive members separated by twice the rotational constant. As the field strength is increased, the transitions involving the lowest J values shift rapidly to higher frequencies. Those with higher J values are less sensitive to the field, by virtue of gyroscopic

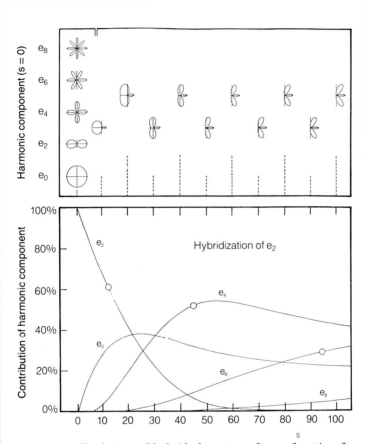

FIGURE 13. Evolution of hybrid character of wavefunction for $\psi(e_2)$ state with increase of s-parameter ($s \equiv 8\omega$) corresponding to electric field strength. Lower panel shows percentage contribution of component circular harmonics (the field-free wavefunctions, for $s = 0$); these curves show normalized values of the squares of amplitude coefficients in the hybrid wavefunction. Points located on e_2, e_4, and e_6 curves indicate s-value for which the eigenstate first becomes bound in the cosine potential. Upper panel shows polar plots of probability distributions, as in Fig. 12.

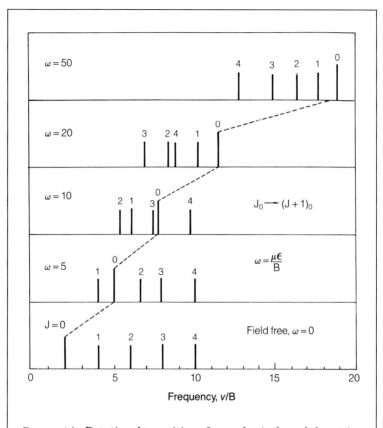

FIGURE 14. Rotational transitions for a spherical pendulum/pin-wheel (transition frequency ν in units of the rotational constant, B), derived from tables of K. von Meyenn, *Z. Phys.* **231**, 154 (1970). For simplicity, only a few members of one series of transitions, $J_M \to (J+1)_M$ is shown, with $M = 0$ and the nominal $J = 0$ to 4. Many other series exist since, although $\Delta M = 0$ or ± 1 still holds, ΔJ is not limited to ± 1 due to field-induced hybridization of the J-states. Lowest spectra pertains to field-free case; others to increasing values of $\omega = \mu\varepsilon/B$.

stability, but respond in the same way when the field becomes sufficiently high. Accordingly, as the field is scanned, the spectral pattern "turns inside-out"; as in some folk dances, the lower J transitions promenade through the array of higher J lines, until at sufficiently high field the frequency sequence of the transitions is reversed. This reflects the J-hybridization of the states and the conversion from pinwheel rotation to pendular libration.

In current experiments, Bretislav Friedrich is using laser-induced fluorescence from the ICl molecule to observe the shifts of transition frequencies and intensities as a function of the imposed ε-field. In this way, he can obtain direct evidence for the trapping of the molecular dipole in pendular states. The field-dependence of the Stark shifts provides a measure of the extent of orientation achieved, since the expectation value of the direction cosine is given by

$$\langle \cos \theta \rangle = -\partial(E/B)/\partial\omega, \tag{3}$$

according to the Hellmann–Feynman theorem. Likewise, incisive tests are available by virtue of the drastic hybridization produced by the high field. The mixing of J-states will enrich the spectra with many transitions that would be forbidden in the absence of the field.

The spectroscopy of oriented and "highly hybridized" molecules offers many other useful features. For instance, the ICl fluorescence Bretislav is studying comes from a vibrational band ($v' = 19$) of the $A\ ^3\Pi$ electronic state in which the dipole moment is similar in magnitude but is suspected to be opposite in sign to that for the ground state. Reversing the sign of the dipole is equivalent to shifting the potential pictured in Fig. 11 by $\theta \to \theta + 180°$. If, for example, the laser-induced excitation from the ground electronic state $X\ ^1\Sigma$ occurs with $\theta \sim 0°$, this will correspond to the most attractive region of the potential well for the ground state but to the most repulsive region for the excited A state. Accordingly, the transition probability for $0_0(X) \to 1_0(A)$ will depend on the exponentially decaying tail of the $1_0(A)$ wavefunction in the classically excluded region, once ω becomes large enough to ensure that both $0_0(X)$ and $1_0(A)$ are pendular states. This will make such transitions extremely sensitive to the librational amplitude. Orienting molecules by rotational hybridization, an approach provoked by a Garden of Eden problem, now seems likely to yield a bountiful harvest.

Dimensional Scaling of Electronic Tunneling: Resonance Redux

The formidable difficulty of computing electronic structures with accuracy comparable to experiment is notorious. In a recent review, Karplus noted that over the past 40 years computing power has increased by a factor of $\sim 10^{10}$ (of which 10^7 is attributed to improved hardware, 10^3 to software). Yet during this period the number of electrons that can be treated with definitive accuracy has increased at most by 10-fold, from two to 20 or so. At present, the major approach for quantitative calculations remains the Hartree–Fock (HF) self-consistent field approximation, augmented by configuration interaction (CI). The correlation energy, defined as the error in the HF approximation, is typically comparable to chemical bond strengths or activation energies and has proved extremely recalcitrant. Often CI calculations with even millions of configurations are inadequate. Despite these daunting aspects, current electronic structure calculations have become extremely useful tools. Chemists, like archaeologists, know how to make the most of any available evidence.

My group has been exploring a dimensional scaling method which offers new computational strategies and heuristic perspectives. This method allows major features to be evaluated by purely classical or semiclassical means. Simple leading approximations in the large-dimension limit correspond to the prequantum notions of Lewis and Langmuir. For atoms with up to 100 electrons, these leading terms give ground-state energies comparable to the Hartree–Fock method but require very little computation and involve an entirely different physical picture. For two-electron atoms, energies accurate to one part in 10^6 or better are readily obtained by dimensional interpolation or perturbation expansions; remarkably, the correlation energy is nearly a linear function of the reciprocal of the spatial dimension. In this approach, Pauling's concept of resonance emerges quite naturally in terms of tunneling among equivalent Lewis structures. There is no need to invoke contaminating assumptions about the form of the wave-function. Recently, for the prototype case of H_2^+, we have found that the exchange energy due to resonance in the actual 3D molecule can be accurately computed from a *classical* electrostatic potential evaluated quite simply at the large-D limit. This is a striking result, since electron exchange and tunneling have been regarded as quintessential quantum phenomena. The article on resonance in *Chemical Reviews* which Linus wrote at the age of 27 has long been cited as classic; from the perspective of dimensional scaling that approbation now seems doubly deserved.

Taking the dimension of space as a variable has proven a useful expedient for many problems not amenable to ordinary methods, especially in statistical mechanics, particle and nuclear physics, and quantum optics. Typically the problem is solved analytically for some "unphysical" dimension $D \neq 3$ where the physics becomes much simpler, and perturbation theory is employed to obtain an approximate result for $D = 3$. Most often the analytic solution is obtained in the $D \to \infty$ limit, and $1/D$ is used as the perturbation parameter. I was only dimly aware of such stratagems until the fall of 1981, when I noticed a tutorial article in *Physics Today* about quarks, gluons, and "impossible problems" of quantum chromodynamics. The author, Edward Witten, illustrated the $D \to \infty$ limit with a rough calculation for helium. His result for the ground state energy was off by 40%, pitiful compared with the 5% accuracy of conventional first-order perturbation theory or the 1.5% accuracy of the Hartree–Fock approximation. However, since I was teaching a quantum mechanics course and on the lookout for provocative problems, I tried setting up the helium example for a homework exercise. Soon I found that if the large-D limit was recast a bit, a very simple calculation gave 1% accuracy. This encouraged me to try to make use of another unphysical limit, $D \to 1$, which had a known solution. In order to interpolate, I assumed for the D-dependence a geometric series in powers of $1/D$, fixing the parameters by means of the simple, exactly calculable $D \to \infty$ and $D \to 1$ limits. Plugging $D = 3$ into the result of this easy exercise gave the correct energy within two parts in 10^5. Exciting and mysterious! Of course this could have been a deceptive coincidence, but it obliged me to find out.

The pursuit of dimensional interpolation and scaling over the past decade has captivated some very able students and led us to other intriguing surprises. Here I will briefly outline the method as background for our recent study of resonance; several reviews of other results are now available and a tutorial volume is forthcoming.

The starting point is to generalize the Schrödinger equation to D-dimensions. This is done by simply endowing all vectors with D Cartesian components. Figure 15 illustrates the procedure for polar coordinates. Thereby the Laplacian operator in the kinetic energy and the Jacobian volume element are modified, but the potential energy retains the same form as for $D = 3$. Furthermore, the D-dependence can readily be removed from the Laplacian and the Jacobian J by setting

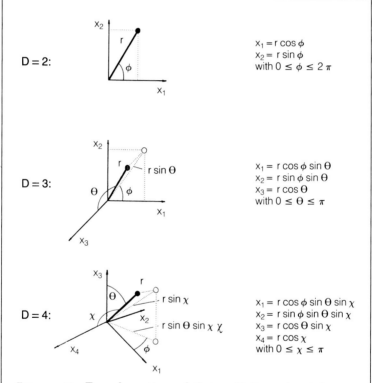

FIGURE 15. Transformations defining D-dimensional hyperspherical polar coordinates in terms of Cartesian coordinates, illustrated for $D = 2, 3, 4, \ldots$. On going from the D to $D + 1$ case, a further Cartesian axis x_{D+1} is added; the radius vector $\mathbf{r}$ is then projected on this axis via the cosine of the new polar angle θ_{D+1} and projected on the D-dimensional subspace via the sine of that angle.

$\Psi = J^{-1/2}\Phi$ and solving, not for the wavefunction Ψ, but for Φ, the square root of the probability distribution function. Such a transformation gives rise to a centrifugal potential U, which contains all the explicit D-dependence, simply a quadratic factor. Next we introduce dimension-scaled distance units, proportional to the radius of the maximum in the probability distribution for the hydrogen atom. This inflates the distance scale by a factor of $(D - 1)^2$ and compresses the energy scale by $(D - 1)^{-2}$. In absolute units, the energy vanishes as $D \to \infty$ and becomes singular as $D \to 1$, but the scaled energy is finite in both these limits.

For large D, the rescaled Schrödinger equation takes a perspicuous form. The centrifugal potential U and the Coulombic potential V become independent of dimension, whereas the remaining kinetic energy terms scale as $1/D^2$. In effect, the factor $\hbar^2/m_e$ involving Planck's constant and the electronic mass, which occurs in the unscaled kinetic energy, is replaced by $1/D^2$ in the scaled version, whereas this factor cancels from the scaled centrifugal potential. The limit $D \to \infty$ is thus tantamount to $\hbar \to 0$ and/or $m_e \to \infty$ in the kinetic energy. The electrons then assume fixed positions relative to the nuclei and each other. This electronic geometry corresponds to the minimum of the effective potential, $W = U + V$, the sum of the rescaled centrifugal and Coulombic terms. We call this the *Lewis structure;* it can be calculated exactly from classical electrostatics for any atom or molecule and provides a rigorous version of the qualitative electron-dot formulas introduced by Gilbert Newton Lewis in 1916.

For D finite but very large, equivalent to a very heavy electronic mass, the electrons are confined to harmonic oscillations about the fixed positions attained in the $D \to \infty$ limit. We call these motions *Langmuir vibrations,* to acknowledge the prescient suggestion by Irving Langmuir in 1919 that "the electrons could . . . rotate, revolve, or oscillate about definite positions in the atom." In the dimensional perturbation treatment the first-order term, proportional to $1/D$, corresponds to these harmonic vibrations, so the coefficient of this term is calculable from the curvatures of the effective potential about its minimum. Accordingly, the standard methods for normal mode analysis of molecular vibrations devised by Bright Wilson become directly applicable to electronic structure.

The large D limit obtained from dimensional scaling may be termed *pseudoclassical*, to indicate that it is not the same as the conventional classical limit obtained by $\hbar \to 0$ for fixed dimension. Since the unscaled centrifugal potential is proportional to $\hbar^2$, it does not contribute to the conventional classical limit. With dimensional scaling, however, the centrifugal term introduces barriers which prevent the electrons from falling into the nucleus and cures other ills of the old Bohr–Sommerfeld quantum theory, thereby fostering use of modern semiclassical methods.

As D decreases further, equivalent to a lighter electronic mass, the electron oscillations become increasingly anharmonic and change character from semiclassical to semiquantal. Eventually, for low D there occur the wild excursions corresponding to a strongly quantal domain. Figure 16 illustrates for the hydrogen atom this progression in the radial probability distribution function. For two-electron atoms, instructive aspects of the D-dependence have been mapped out by John Loeser, Doug Doren, and David Goodson, my first varidimensional graduate students. Among other things, their work established that the dimension dependence is dominated by singularities that occur at the $D \to 1$ limit. This limit is tantamount to $\hbar \to \infty$ in the unscaled Schrödinger equation, and hence represents a *hyperquantum* limit. The singular terms at $D = 1$ can be readily evaluated to any desired accuracy, so the contribution of these terms can be deducted. The remaining dimension dependence is then mild and can be well approximated by the semiclassical dimensional perturbation expansion in powers of $1/D$ about the $D \to \infty$ limit. For instance, simply combining two first-order perturbation calculations, one performed at the $D \to 1$ limit, the other at the $D \to \infty$ limit, yields 90% of the correlation energy for helium. That is far easier than a comparable configuration interaction calculation.

A remarkably simple and effective treatment of many-electron atoms has been given by John Loeser, now on the faculty at Oregon State, Linus' alma mater. He evaluated the Lewis + Langmuir terms, corresponding to a first-order dimensional perturbation, with the approximation that at $D \to \infty$ the electrons are equidistant from each other and equidistant from the nucleus. The minimum that defines the Lewis structure can then be easily found (as the smallest positive root

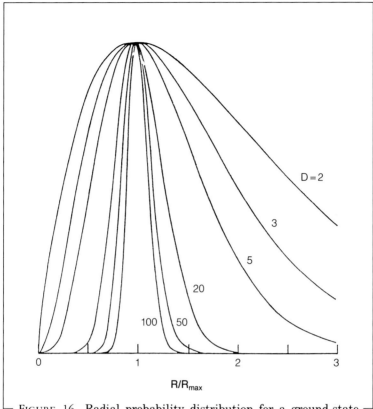

FIGURE 16. Radial probability distribution for a ground-state hydrogenic atom in D-dimensional space. The curves (labeled with values of D) are normalized at the maximum, which occurs at $R_{max} = [\frac{1}{2}(D - 1)]^2/Z$.

of a quartic equation). This puts the electrons at the corners of a regular N-point simplex, a "hypertetrahedron," while the nucleus lies along an axis that passes perpendicularly through the centroid. Surely G. N. Lewis would be delighted at this generalization of his cubical atom! With some further approximations, Loeser obtained total energies with maximum errors of only about 1%, for atoms with up to $N \sim 100$ electrons. This is comparable to but vastly easier to compute than single-ζ Hartree–Fock results.

In applications of dimensional scaling to molecules, pursued by Don

Frantz and other students in my group, a new feature appears that corresponds to *Pauling resonance*. The simplest examples are shown in Fig. 17, which depicts the effective potentials at the $D \to \infty$ limit for the H_2^+ and H_2 molecules. At the united atom limit, these systems become He^+ and He, respectively, and the effective potential has a single minimum. As the internuclear distance R is increased, however, mul-

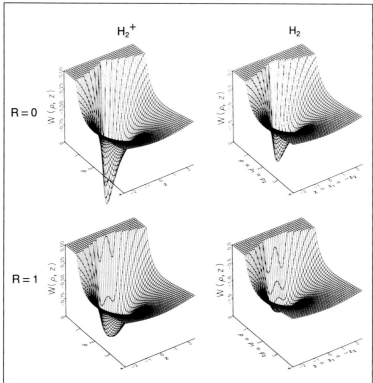

FIGURE 17. Effective potential surfaces $W = U + V$ for the H_2^+ and H_2 molecules for $D \to \infty$ and various internuclear distances R (in bohr units). The united atom limit at $R = 0$ pertains to He^+ and He, respectively, and exhibits a single minimum. For H_2^+, when R increases above a critical value, $R_c \approx 1.2990$, a double minimum becomes increasingly pronounced. For H_2, when R exceeds ≈ 0.9111 an analogous double minimum develops. The H_2 surface depends on five coordinates (*cf.* Fig. 18), so

tiple minima develop. For instance, the Lewis structure for H_2 when R is small has both electrons in the plane bisecting the molecular axis, but when R becomes large enough the effective potential acquires two pairs of double minima. One pair corresponds to localizing each electron on a different nucleus; the other pair, much less favorable energetically, has both electrons on one or the other nucleus.

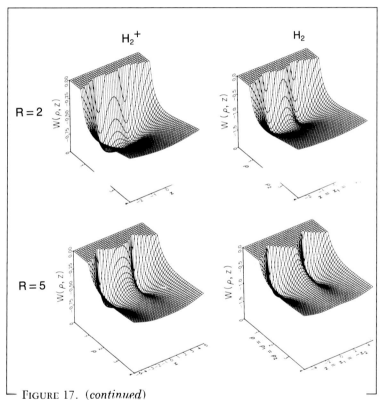

FIGURE 17. (*continued*)
for this plot three constraints are imposed: The electrons reside on a cylinder of radius ρ coaxial with the molecular axis ($\rho_1 = \rho_2$), at equal distances above and below the plane bisecting the nuclei ($z_1 = -z_2$), and with the dihedral angle f held constant at the value obtained at the minima. Not evident are a secondary pair of minima (with $z_1 = z_2$) in which both electrons are above or below the bisector plane (*cf.* Figs. 18 and 19).

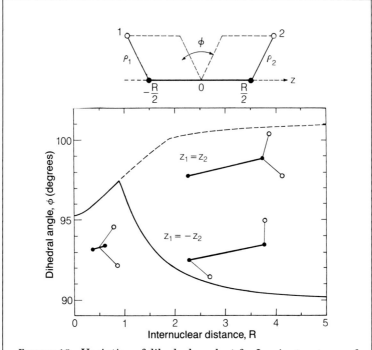

FIGURE 18. Variation of dihedral angle ϕ for Lewis structures of hydrogen molecule with internuclear distance R (bohr units). Throughout, radial distance of electrons from molecular axis ($\rho_1 = \rho_2 = \rho_m$) corresponds to minimum of effective potential (*cf.* Fig. 17). For small R, electrons lie in the plane bisecting the molecular axis and the dihedral angle is close to the value 95.3° pertaining to the united atom limit. Symmetry breaking occurs at two critical points: $R_c \approx 0.9111$, where $\phi_m \approx 97.51°$ and $\rho_m = 0.9195$; and $R_c \approx 1.9137$, where $\phi_m \approx 100.14°$ and $\rho_m = 1.3532$. Stick figures show typical structures at smaller and larger R.

Figure 18 illustrates these "electronic isomers" by showing the dependence on R of the dihedral angle ϕ between two planes hinged on the molecular axis, each containing one of the electrons. When viewed along the molecular axis, for small R the Lewis structure resembles that for the helium atom and ϕ is near 95°. The atom is bent rather than linear because the angle between the electron radii at the potential minimum reflects the competition between centrifugal repulsion (minimal for 90°) and interelectron repulsion (minimal for 180°). As R

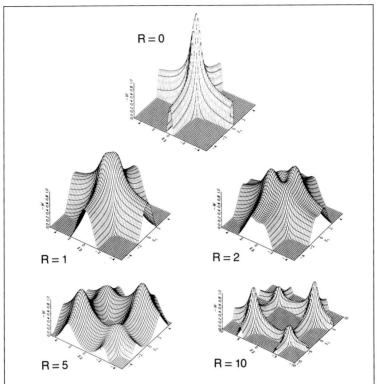

FIGURE 19. Dependence on internuclear distance R of the H_2 effective potential surface as functions of the individual linear coordinates z_1 and z_2 at constant ρ and ϕ (*cf.* Fig. 18). Negative energies are plotted to better exhibit changes in the minima as R varies. The surface is symmetric in z until the first splitting occurs at $R_c \approx 0.9111$, which forms a pair of global minima with $z_1 = -z_2$; the second splitting at $R_c \approx 1.9137$ forms another pair of minima with $z_1 = z_2$.

increases toward a critical value, the optimum dihedral angle opens up by a few degrees. At larger R, the dihedral angle decreases rapidly towards 90° for the most favorable structure (peroxide-like) but opens up further for the less favorable one (amino-like). Thus, in the $D \to \infty$ limit the effective potential changes form as R varies, exhibiting minima that correspond to a different Lewis structure for each distinct valence bond configuration. Figure 19 shows another plot of the H_2 effective potential which brings out this aspect. As D is decreased, equivalent to lowering the electronic mass, tunneling of electrons among these various minima in the effective potential becomes increasingly important; this corresponds to resonance among the different valence bond structures.

For the H_2^+ molecule–ion, the Schrödinger equation for fixed nuclei can be solved exactly in any dimension, since it is separable in spheroidal coordinates. By virtue of the inherent hypercylindrical symmetry, an exact interdimensional degeneracy links the D-dependence to the orbital angular momentum projection on the internuclear axis. Each unit increment in the projection quantum number $|m|$ corresponds to inflating the dimension by two steps, $D \to D + 2$. Consequently, for any odd D the energies and wavefunctions can be derived by suitably scaling excited states of the $D = 3$ molecule. Figure 20 shows the electronic energy for several such H_2^+ states. The gerade, ungerade pairs for $|m| = 0, 1, 2, \ldots$ (i.e., $\sigma, \pi, \delta, \ldots$) correspond to hydrogenic ground-states for $D = 3, 5, 7, \ldots$ in the separated atom limit, $R \to \infty$. The splitting ΔE_D between these pairs (shown shaded), grows exponentially as the atoms approach. As discussed by Linus in his 1928 paper, this splitting represents the resonance energy arising from exchange of the electron between the protons.

Using the H_2^+ double minimum potential for the large-D limit, shown in Fig. 17, Sabre Kais has determined the resonance energy in the most rudimentary fashion, as the energy splitting ΔE_D produced by electronic tunneling between the minima. This unorthodox approach proves to give good agreement with the exact numerical calculations of Frantz for a wide range of D and R. A key aspect is a scaling law for the splitting, a consequence of the interdimensional degeneracy. Figure 21 illustrates this scaling, which permits the treatment to be simplified by using any convenient D, in particular the $D \to \infty$ limit.

The tunnel effect splittings are computed using the instanton method, developed particularly by William Miller, another scientific grandson of Linus by way of Bright Wilson. This method examines the evolution of the system in *imaginary* time, equivalent to motion in real time in an inverted potential like that of Fig. 19. In this way, the classically forbidden trajectories under the actual potential barrier become allowed trajectories in the inverted potential. To leading order in

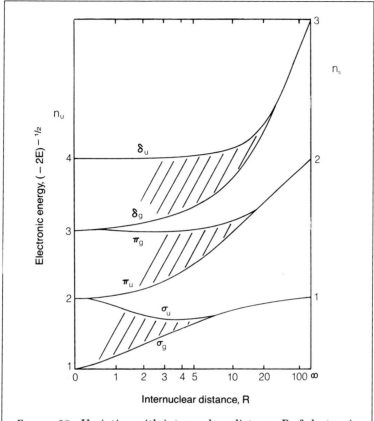

FIGURE 20. Variation with internuclear distance R of electronic energy for several states of the H_2^+ molecule–ion (omitting nuclear repulsion). Abscissa scale is $10R/(R + 5)$ in order to cover full range from united atom limit $(R \to 0)$ to separated atom limit $(R \to \infty)$. Ordinate scale is $(-2E)^{-1/2}$ so that limiting values correspond to hydrogenic principal quantum numbers. The pairs of gerade, ungerade states shown stem from separated atom states with the maximum projection of the electronic angular momentum $m = n_s - 1$, for $n_s = 1, 2, 3, \ldots$ in the $D = 3$ molecule. These states are related by an exact interdimensional degeneracy and thus correspond to $m = 0$, $n_s = 1$ for $D = 3, 5, 7, \ldots$, respectively. The tunnel effect splitting ΔE_D (shown shaded) between the pairs of g, u states represents the exchange or resonance energy associated with switching the electron between the protons.

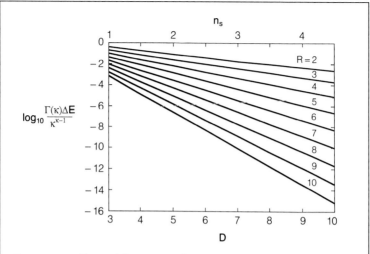

FIGURE 21. Test of dimensional scaling for tunnel effect splittings or resonance energies of H_2^+, for internuclear distance $R = 2, 3, \ldots 10$ (bohr units). Here $\kappa = (D - 1)/2$. Splittings obtained from exact numerical calculations of Frantz are weighted with D-dependent factor and found to vary linearly with D, in accord with scaling inferred from analysis by Kais of large-D regime.

Planck's constant, the energy splitting is given by

$$\Delta E = A \exp(-S_0/\hbar), \qquad (4)$$

where S_0 is the classical action integral for the trajectory of zero-energy between the two maxima in the inverted potential. This trajectory is called the instanton path. The prefactor A involves contributions from fluctuations about the instanton path. Even for H_2^+ these functions S_0 and A are not easy to calculate directly for $D = 3$, except for large R.

The strategem of calculating in the large-D regime and then scaling to get the $D = 3$ result introduces remarkable simplicity. The zero-energy trajectory corresponds to traveling peak-to-peak in the inverted potential, or from the bottom of one potential well to the other in the actual potential; that is just where the electrons reside in the $D \to \infty$ limit. Figure 22 shows the instanton trajectories for the tunneling paths

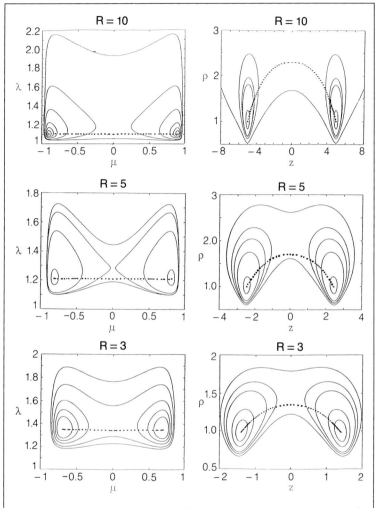

FIGURE 22. Instanton paths for tunneling between pairs of minima in the effective potential surface at $D \to \infty$ for the $H_2{}^+$ molecule–ion with $R = 3, 5,$ and 10 bohr radii. Panels at left show contour maps of effective potential in separable form obtained using spheroidal coordinates; at right are maps for non-separable form in cylindrical coordinates.

at three values of R, drawn on contour maps of the double-minimum potential. Maps are shown both for the separable version expressed in spheroidal coordinates (λ, μ) and the nonseparable version in cylindrical coordinates (ρ, z). In the spheroidal version, the double minimum potential occurs in only one coordinate, $\mu = (r_a - r_b)/R$, where r_a and r_b locate the electron relative to the two protons. The tunneling trajectory is then just a straight line. The instanton action becomes an easy integral, given by

$$S_0 = [2a/(1 - a^2)] - \ln[(1 + a)/(1 - a)], \tag{5}$$

where $a \equiv \mu_m(R)$ specifies the position of the potential minima; the fluctuation factor is given by

$$A = 4(2\hbar\omega V_0/\pi)^{1/2}, \tag{6}$$

where $\omega = 2a/(1 - a^2)^2$ is the frequency for harmonic vibrations within the minima and $V_0 = \omega a^3/4$ is the height of the barrier (at $\mu = 0$) between the minima. It is gratifying that such extremely simple, analytic results are quite accurate. Our devious route to such results—pursuing trajectories both in imaginary time and in unphysical dimensions—may, we hope, lead to more effective treatments of multielectron systems.

Although our unorthodox route to resonance in H_2^+ is devious, it is also rigorous. Despite the great utility and intuitive appeal of the resonance concept as advocated by Pauling, it has drawn many polemic attacks. In particular, critics have repeatedly pointed out that the exchange or resonance integrals that appear naturally in a primitive valence bond treatment of H_2^+ or H_2, such as Linus gave in his 1928 paper, do not appear in a more rigorous treatment. By extension, these critics conclude that for any molecule resonance is an artifact, a misleading consequence of an inadequate wavefunction. Thus it is satisfying that the large-D limit, which involves no assumptions about the wavefunction, exhibits Pauling resonance so clearly, in terms of the tunneling of electrons among the various effective potential minima corresponding to distinct Lewis structures.

The surprising effectiveness of dimensional scaling suggests that it may substantially enhance both *ab initio* calculations and qualitative conceptual analysis of chemical bonding and reactivity. This remains to be seen, as thus far we have examined only a few of the simplest molecules, and chiefly just in the large-D regime. Since the significant variable is $1/D$, the very simple pseudoclassical $D \to \infty$ limit is closer to the "real world" (at $1/D = 1/3$) than is the less simple hyperquantum $D \to 1$ limit. In both limits all electron correlation is included. The ability to approximate the far more difficult $D = 3$ solution by means of dimensional interpolation or extrapolation hence does not depend on the magnitude of the electronic interactions but only on the dimension dependence. It is heartening that the D-dependence has proven to be simple and generic for our prototype examples. This holds also for a kindred correlation problem, the virial coefficients of hard sphere fluids.

For heuristic analysis, the pseudoclassical character of the large-D regime is a great advantage. It also addresses a persistent philosophical question of quantum theory: Might not "hidden" classical variables exist? At first blush, the fixed electronic geometry for Lewis structures and defined modes for Langmuir vibrations appear to violate the uncertainty principle. Computing electronic tunneling in H_2^+ from classical trajectories on a classical electrostatic potential seems even more egregious. However, dimensional scaling of the coordinates implies that the conjugate momenta are scaled inversely, so the commutators and the uncertainty principle remain invariant. In effect, we transform to a strange space which brings out reticent classical structure and *hides the quantum mechanics*. This allows our seemingly classical calculations of the Lewis and Langmuir terms and Pauling resonance energies.

Like my first visit to Linus Pauling 33 years ago, compiling this chapter has made me realize more fully the awesome impact and scope of his ideas. But even more I want to pay homage to his joyful, steadfast devotion to science and humanity. To know him is exhilarating for many reasons, but especially because he has taught me and so many others to look at this world, as Lewis Carroll said of Alice, with "a pure unclouded brow and dreaming eyes of wonder."

ACKNOWLEDGMENTS

I am deeply grateful for the privilege of pursuing science in collaboration with many able students and colleagues. Also vital has been support received over the years from several agencies, chiefly the National Science Foundation. I am especially thankful to Don Braben and his Venture Research Unit for making possible our recent work on dimensional scaling when it had been dismissed by other agencies as too quixotic.

REFERENCES

For each topic, some key reviews and a few papers treating specific aspects are cited.

REACTION DYNAMICS:

Herschbach, D. R. (1987). "Molecular Dynamics of Elementary Chemical Reactions," *Angew. Chem. Int'l. Ed. Engl.* **26**, 1221.

Karplus, M. (1968). "Structural Implications of Reaction Kinetics," in *Structural Chemistry and Molecular Biology: A Volume Dedicated to Linus Pauling* (A. Rich and N. Davidson, eds.), pp. 837–847. W. H. Freeman and Co., San Francisco.

King, D. L., and Herschbach, D. R. (1973). "Facile Four-Centre Exchange Reactions," *Chem. Soc. Faraday Disc.* **55**, 331.

Kwei, G. H., and Herschbach, D. R. (1969). "Molecular Beam Kinetics: Reactions of Alkali Atoms with ICl and IBr," *J. Chem. Phys.* **51**, 1742.

Lee, Y. T. (1987). "Molecular Beam Studies of Elementary Chemical Processes," *Science* **236**, 793.

Levine, R. D., and Bernstein, R. B. (1987). *Molecular Reaction Dynamics and Chemical Reactivity.* Oxford Univ. Press, New York.

Polanyi, J. C. (1987). "Some Concepts in Reaction Dynamics," *Science* **236**, 680.

ORIENTING MOLECULES:

Barnwell, J. D., Loeser, J. G., and Herschbach, D. R. (1983). "Angular Correlations in Chemical Reactions," *J. Phys. Chem.* **87**, 2781.

Bernstein, R. B., Herschbach, D. R., and Levine, R. D. (1987). "Dynamical Aspects of Stereochemistry," *J. Phys. Chem.* **91**, 5365.

Friedrich, B., and Herschbach, D. R. (1991). "Spatial Orientation of Molecules in Strong Electric Fields: Evidence for Pendular States," *Nature* **353**, 412.

Friedrich, B., and Herschbach, D. R. (1991). "On the Possibility of Orienting Rotationally Cooled Polar Molecules in an Electric Field," *Z. Phys.* **D18**, 153.

Friedrich, B., Pullman, D. P., and Herschbach, D. R. (1991). "Alignment and Orientation of Rotationally Cool Molecules," *J. Phys. Chem.* **95**, 8118.

Kim, S. K., and Herschbach, D. R. (1987). "Angular Momentum Disposal in Atom Exchange Reactions," *Chem. Soc. Faraday Disc.* **84**, 159.

Loesch, H. J., and Remscheid, A. (1990). "Brute Force in Molecular Reaction Dynamics: A Novel Technique for Measuring Steric Effects," *J. Chem. Phys.* **93**, 4779.

Pauling, L. (1930). "The Rotation of Molecules in Crystals," *Phys. Rev.* **36**, 430.

DIMENSIONAL SCALING:

Doren, D. J., and Herschbach, D. R. (1987). "Two-Electron Atoms Near the One-Dimensional Limit," *J. Chem. Phys.* **87**, 433.

Frantz, D. D., and Herschbach, D. R. (1988). "Lewis Electronic Structures as the Large-Dimension Limit for H_2^+ and H_2 Molecules," *Chem. Phys.* **126**, 59.

Frantz, D. D., and Herschbach, D. R. (1990). "Interdimensional Degeneracy and Symmetry Breaking in D-Dimensional H_2^+," *J. Chem. Phys.* **92**, 6668.

Goodson, D. Z., and Herschbach, D. R. (1987). "Electron Correlation Calibrated at the Large-Dimension Limit," *J. Chem. Phys.* **86,** 4997.

Herschbach, D. R. (1987). "New Dimensions in Reaction Dynamics and Electronic Structure," *Chem. Soc. Faraday Discussion* **84,** 465.

Herschbach, D. R., Avery, J., and Goscinski, O., eds. (1992). *Dimensional Scaling in Chemical Physics.* Kluwer Academic, Dordrecht, The Netherlands.

Kais, S., Frantz, D. D., and Herschbach, D. R. (1991). "Electronic Tunneling in H_2^+ from Instanton Analysis of the Large-Dimension Limit," *J. Chem. Phys.* **95,** 9028.

Karplus, M. (1990). "Three-Dimensional Pople Diagram," *J. Phys. Chem.* **94,** 5435.

Loeser, J. G. (1987). "Atomic Energies from the Large-Dimension Limit," *J. Chem. Phys.* **86,** 5635.

Loeser, J. G., Zhen, Z., Kais, S., and Herschbach, D. R. (1991). "Dimensional Interpolation of Hard Sphere Virial Coefficients," *J. Chem. Phys.* **95,** 4525.

Pauling, L. (1928). "The Application of the Quantum Mechanics to the Structure of the Hydrogen Molecule and the Hydrogen Molecule–Ion and to Related Problems," *Chem. Revs.* **5,** 173.

Witten, E. (1980). "Quarks, Atoms, and the $1/N$ Expansion," *Physics Today* **33**(7), 38.

9

REAL-TIME
LASER FEMTOCHEMISTRY

Viewing the Transition from
Reagents to Products

Ahmed Zewail
California Institute of Technology
and

Richard Bernstein
University of California, Los Angeles

One of the most fundamental problems in chemistry is understanding how chemical reactions occur: that is, how reagents make their journey to products. Traditionally, chemists start by studying the thermodynamics of a reaction, then its rate, and finally postulate its mechanism.

The century old and well-known Arrhenius rate equation

$$k(T) = Ae^{-E_a/kT}$$

systemizes a large body of experimental data, expressing a reaction's rate in terms of the activation energy, E_a, and the pre-exponential A-factor. For gas-phase bimolecular reactions, the equation conveys the essence of the reactive encounter: Molecules collide (expressed by the A-factor, which is related to the rate constant for collisions), but they react only if collisions are sufficiently energetic (the exponential term).

The rate constant, $k(T)$, however, does not provide a detailed molecular picture of the reaction. This is because $k(T)$ is an average of the microscopic, reagent-state to product-state rate coefficients over all possible encounters. These might include different relative velocities, mutual orientations, vibrational and rotational phases, and impact parameters. We needed a way to describe, with some quantitative measure, the process itself of chemical reaction: how reagent molecules approach, collide, exchange energy, sometimes break bonds and make new ones, and finally separate into products. Such a description is the goal of molecular reaction dynamics.

Some 30 years old now, molecular reaction dynamics has matured into a major field of chemistry. It has had impact on photochemistry and laser chemistry; laser mass spectrometry and ultrasensitive detection; isotope separation; energy transfer and relaxation in gases and solutions; disequilibrium and transport phenomena; atmospheric, combustion, and plasma chemistry; dynamics of gas–solid interactions and heterogeneous catalysis; cluster formation and cluster chemistry; and even dynamic photobiology.

The Pimentel Report, "Opportunities in Chemistry," stresses chemical reaction dynamics as one of the most important current research areas in chemistry. As further evidence of the significance of the field, the 1986 Nobel Prize in Chemistry was awarded to Dudley R. Herschbach, Yuan T. Lee, and John C. Polanyi for their incisive experimental and interpretive work on the dynamics of elementary chemical reactions.

Major advances in the experimental study of molecular reaction dynamics have come from the application of molecular beam and laser techniques. In the simplest molecular beam experiments, a beam of reagent molecules, say A, is directed toward coreagent molecules B (in the form of a target gas or another beam), and the reactive scattering that produces product molecules C and D is observed. The relative kinetic energy of the reagents can be changed in these "single collision" experiments by varying the velocity of A with respect to B.

For laser–molecular beam experiments, a laser excites one of the reagent molecules and thus influences the reaction probability or initiates a unimolecular process by energy deposition in a molecule, say ABC. In this so-called half-collision process, the fragmentation of the

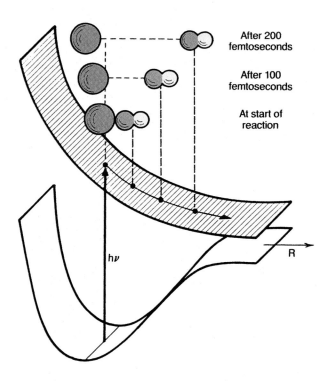

Photofragmentation of a triatomic molecule occurs in femtoseconds

After 200 femtoseconds

After 100 femtoseconds

At start of reaction

hν

R

Photofragmentation of a triatomic molecule such as ICN is shown schematically here. The lower surface represents a strip of the ground state potential energy surface as a function only of R, the separation between the atom and the diatomic fragment. The absorption of a photon (energy hν) excites the triatomic molecule from its ground state to its upper, repulsive potential energy surface at the start of the reaction. After a few hundred femtoseconds, the photofragments are essentially free of one another. Experiments can measure the real-time progress of the reaction as the triatomic reactant makes the transition to its diatomic and atomic products. Since the recoil velocity of fragments is typically 1 km per second, in 100 fs the distance spanned is 1 Å. [Adapted from M. Rosker, M. Dantus, and A. H. Zewail, *Science,* **241**, 1200 (1988).]

excited ABC molecule into AB and C is the dynamical process studied. In such experiments, the observations are of the angular, velocity, and quantum state population distributions of the products.

The quantum state distribution of product molecules can be obtained using infrared or visible chemiluminescence, laser-induced fluorescence, or resonance-enhanced multiphoton ionization. Product angular and velocity distributions are best measured by rotatable mass spectrometer detectors with time-of-flight analysis of products. The task is to relate these observables to the potential energy governing the interaction between molecular or atomic fragments. Measurements of reaction cross-sections and their angular dependence as a function of collision energy are compared with theoretical (usually computational) predictions, based on *ab initio* or semiempirical calculations of potential energy.

At the very least, such studies reveal the strength and range of intermolecular forces. Much more has been learned from detailed collision experiments in which the vibrational and rotational states of both the reactant and product molecules are known. Such "state-to-state" experiments provide information on the role of precollision states in determining the states of the product molecules.

The problem is that a major feature of reaction dynamics is still unseen experimentally, namely the transition state between reagents and products. This state often lasts only picoseconds or less, and experiments on a longer time scale provide data that are effectively time-integrated over the entire course of the collision. As emphasized recently by Ian W. M. Smith of the University of Birmingham, U.K., a great deal is known about the "before" and "after" stages of the reaction but it is difficult to observe the "during" phase.

Now chemical dynamics is becoming more ambitious: It is trying to describe transition-state chemistry. Until recently, this has been approachable only by theoretical studies. The goal now is to make direct, real-time observations of the process of reaction itself, even though the duration of the event may be a matter of only femtoseconds (1 fs is 10^{-15} second).

Femtochemistry, or chemistry on the femtosecond time scale, can be defined as the field of chemical dynamics concerned with the very act of chemical transformation, the process of breaking one chemical bond and making another. On this time scale the molecular dynamics are "frozen out," and the complete evolution of the chemical event is observed. This time scale is perhaps the ultimate one as far as chemistry is concerned, but it is a mistake to infer that new studies will end at this time resolution.

Femtochemistry requires ultrafast laser techniques to initiate and record snapshots of chemical reactions with femtosecond time resolution. It is also advantageous to work with molecules in supersonic beams or expanded jets; the expansion extensively cools the reactant molecules, which simplifies their internal energy distribution and makes it easier to do state-to-state experiments. However, since collisions play no role on the femtosecond time scale, experiments also can be carried out in bulk gases.

The objective of this article is to highlight recent developments in femtochemistry and to make the connection between these real-time ultrafast techniques and earlier methods devoted to studies of reaction dynamics under collisionless conditions. Reaction dynamics in collisional environments, including condensed phases, is beyond our scope.

Real-time femtochemistry of elementary reactions allows reaction dynamics to be viewed in a new light. Because the experiments can utilize laser pulses as short as 6 fs (as of the writing of this article!), the actual progress of the passage through the transition states of a reaction can be viewed directly. For photoinitiated unimolecular and bimolecular reactions, one can obtain snapshots from which the potential relevant to fragment separation and product formation can be deduced with a distance resolution of about 0.1 Å.

We are now seeing a revival of flash photolysis, begun in 1949 by Ronald G. W. Norrish and George Porter at Cambridge University, in the U.K., but the increased nine-orders-of-magnitude improvement in time resolution in femtochemistry allows one to observe the transition states and the fundamental dynamics of chemical bond rupture and bond formation, with subangstrom resolution.

FEMTOCHEMISTRY AND TRANSITION STATES

The critical stage of any reaction is the saddle point separating reagents from products—the crucial transition state configuration. We will use the term "transition states" to encompass all neighboring nuclear configurations important to the transition from reagents to products, including the configuration that defines the transition state in theories of chemical reactions.

Theoretical femtochemistry has been with us since the 1930s, following the introduction of the concept of the transition state by Michael Polanyi in the U.K. and Henry Eyring in the U.S. In a classic work of 1936, Joseph O. Hirschfelder and Eyring, then at Princeton University, calculated point by point the first classical mechanical trajectory of a free hydrogen atom approaching the H_2 molecule to form an "activated complex" or transition state.

In this and all subsequent classical trajectory simulations of elementary chemical reactions, it has been necessary to use numerical integrations of the classical equations of motion with subfemtosecond step sizes in order to accommodate the high-frequency molecular vibrations occurring in the reactants. The hydrogen atoms in H_2 move with a vibrational period of about 7.6 fs; thus, if a single oscillation cycle is to be approximated numerically with more than 10 steps, each step needs to be smaller than 1 fs. In a typical thermal energy atom–molecule collision, a 1 fs time interval corresponds to an atomic displacement of 10^{-2} Å, a step size corresponding to less than 1% of a bond length.

Most of our ideas about the transition state thus far come from such calculations, rather than from experiments. These calculations form the basis for femtosecond transition state chemistry as now envisaged. What is different is that now experimental methods permit observing the transition state and clocking, in real time, the formation and decay of the activated complex. The goal is to observe and monitor the dynamics of bond rupture and bond formation and to measure and characterize the transient intermediate en route to separation into products.

The "entrée" into the transient comes via photons. The tool for this research is femtosecond transition-state spectroscopy of reactions, introduced by the California Institute of Technology group of Ahmed H. Zewail. The combination of ultrashort (picosecond/femtosecond) pulsed laser interferometric and molecular beam techniques, during the past eight years, has made this possible. Pulses as short as 6 fs can now be generated, following the pioneering design of dye lasers by the AT&T Bell Laboratories group of Charles V. Shank, and used in these experiments.

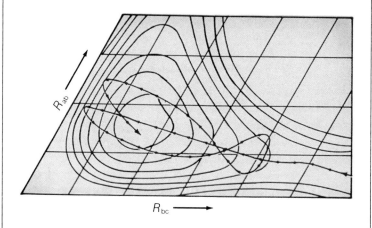

Classical mechanical calculations trace dynamics of a chemical reaction

R_{bc} ——▶

The first classical mechanical trajectory calculation for a chemical reaction, indicated by the curved line with arrows, is plotted on the potential energy surface contour diagram in the skewed coordinate system. The reaction is the collinear collision of H_2 with H: H_aH_b + $H_c \rightarrow H_aH_bH_c \rightarrow H_a$ + H_bH_c. Initially the distance between H_b and H_c (R_{bc}) is very large and R_{ab} is the hydrogen molecule bond length. Note that the trajectory becomes trapped in "Lake Eyring." Although the potential energy surface used for this calculation later turned out to be flawed, the computation is of historical significance. [Adapted from J. O. Hirschfelder, H. Eyring, and B. Topley, *J. Chem. Phys.*, **4**, 170 (1936).]

Traditionally, spectroscopy has been the method of choice for determining the structure of isolated molecules and their intramolecular dynamics. With the advent of intense, ultrashort laser pulses, and the ability to establish accurate time delays in the range 10 to 1,000 fs by micrometer mirror displacements, with auto- and cross-correlation analysis of pulse shapes, it is now possible to do real-time transient spectroscopy of reactions on the femtosecond time scale and thus real-time femtochemistry. Before discussing femtochemistry in detail, we will summarize the key information obtained from time-integrated experimental studies of unimolecular and bimolecular reactions, which will help illustrate the important features of reaction dynamics and the new areas that can be impacted by direct real-time (femtosecond) observations.

BIMOLECULAR REACTIONS

Since the early reactive beam scattering experiments of the 1960s and 1970s, it has become well established that, for a large class of fast elementary reactions involving small molecules, the angular distribution of products with respect to the molecular collision framework—the so-called center-of-mass system—is strongly anisotropic. This asymmetry comes about because the transition state does not last long enough for the activated complex to rotate much, a process that takes, typically, a picosecond. Thus, for many such reactions, called direct-mode, we have experimental evidence of transition states lasting less than a picosecond, based on this crude and average rotational period clock.

On the other hand, for a whole class of complex-mode bimolecular reactions, discovered in 1967 by Herschbach and coworkers of Harvard University, the product angular distributions are symmetrical in the center-of-mass system, implying that the complex sticks together for many rotational periods, typically several picoseconds, before falling apart to yield products. In some cases, they can be stabilized further by subsequent, deactivating collisions, as in unimolecular reactions.

Molecular beam scattering experiments also provide data on angular distributions for elastic scattering, as well as velocity and angular distributions (and their energy dependence) for inelastic scattering. From this information, many features of the anisotropic intermolecular, ground-state, potential surface can be deduced. The accuracy of the derived potential is ascertained by back-calculation of the experimental scattering cross-sections using classical, semiclassical, or (preferably) quantal scattering computations. Classical mechanics, though

Direct-mode reactions take femtoseconds, complex-mode reactions picoseconds

$H_a + H_bH_c \longrightarrow H_aH_b + H_c$

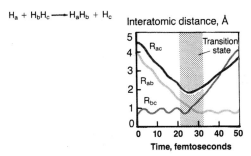

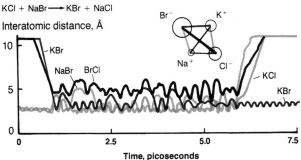

The lifetimes of transition states in bimolecular collisions can vary from several femtoseconds to many picoseconds. The upper graph shows a single trajectory calculated for the direct-mode exchange reaction $H_a + H_bH_c \rightarrow H_aH_b + H_c$, assuming a collinear configuration. Plotted on the ordinate scale are the interatomic separations R_{ab}, R_{bc}, and R_{ca} as a function of time from an arbitrary starting condition. Note the oscillations in R_{bc} before the encounter, then the exchange taking place, forming the new molecule H_aH_b, whose oscillation is seen in R_{ab}. There is a time interval of about 10 femtoseconds during which the system is in a tran-

sition state—a collinear, triatomic $H_aH_bH_c$. The lower graph is a similar portrayal of a single trajectory calculated for the complex-mode, gas-phase metathesis reaction KCl + NaBr→tetraatomic complex (structure shown)→KBr + NaCl. The "snarled trajectories" are evidence for a long-lived (here about 5 picosecond) collision complex that eventually falls apart to yield products. [Upper graph adapted from M. Karplus, R. N. Porter, and R. D. Sharma, *J. Chem. Phys.*, **43**, 3258 (1965), lower from P. Brumer and M. Karplus, *Far. Disc. Chem. Soc.*, **55**, 80 (1973) and P. Brumer, Ph.D. thesis (1972).]

it has some well-known limitations, is very good for visualizing the course of the reaction. For elastic and inelastic scattering, quantal scattering computation has supplanted classical trajectory methods, but for reactive collisions (even simple atom–molecule exchange reactions), quantal calculations are still difficult and costly, so quasiclassical methods are widely used. In recent years, however, great progress has been made in implementation of accurate quantum scattering computations on elementary reactions, such as $H + H_2$.

To obtain reaction cross-sections, one must computer-simulate many trajectories and use Monte Carlo methods of averaging over impact parameter, initial orientation, and vibrational and rotational phases, as pioneered by Martin Karplus, then at Columbia University; the late Donald Bunker, University of California, Irvine; John C. Polanyi at the University of Toronto; and others. However, only a modest number of well-chosen trajectories are needed to visualize the key dynamic features of a reaction, especially the formation and decay of the activated complex. "Motion pictures" of such trajectories have been made by combining many successive snapshots of atomic locations calculated as a function of time. To characterize faithfully the fast vibrations of molecules requires that the snapshots be made at very close time intervals. The time scale of interest for direct-mode reactions ranges from about 50 fs to several picoseconds. It is much longer, of course, for complex-mode reactions.

This great progress in obtaining and simulating product angular distributions and energy-dependent reaction cross-sections enables one to advance some concepts regarding energy disposal and consumption, but these time-integrated observables still do not offer a direct view of transition states themselves.

PHOTOFRAGMENTATION REACTIONS

In a full collision, two molecules approach one another, may or may not exchange energy, and may or may not break apart as new product molecules. The process can be divided into two parts, the incoming and the outgoing half-collisions. Photofragmentation is a classic half-collision process in which a stable molecule is photoexcited to an unstable electronic state in a time that is short compared with any significant nuclear motion. Following the excitation, the two fragments separate and accelerate to a terminal relative velocity governed by the available energy for translation.

Since the classic work on photodissociation in 1934 by the group of Alexander N. Terenin at the University of Leningrad, there have been many important technological advances. State-to-state dynamics of photodissociation reactions have now been studied with the help of lasers and techniques such as photofragment time-of-flight spectroscopy. The pioneering theoretical and experimental work of Richard N. Zare of Stanford University, Richard Bersohn at Columbia University, and Kent R. Wilson at the University of California, San Diego, made it possible to study the correlation between laser polarization and the direction and speed of product recoil, thereby providing knowledge of the process of fragment separation. The new and important information on the dynamics of a variety of photodissociation reactions obtained from these experiments permits deduction of the time-integrated population distribution and angular distribution of the photofragment product states.

Such angular distribution experiments also provide a method for inferring the dissociation lifetime of the laser-excited molecule ABC. Basically, if the molecule dissociates before it has time to rotate significantly, then the original alignment of the excited molecule induced by the laser is preserved and the fragments will be ejected in certain well-defined directions. On the other hand, if the dissociation is very much slower than the average rotational period, the photoproducts will have an isotropic angular distribution. The dissociation lifetime can be estimated from the degree of anisotropy of the observed photofragment angular distribution. The technique is similar to the use of the deviation from symmetry in product angular distributions in crossed-beam experiments to estimate the lifetime of the transition state. Many recent advances have been made in this area, as illustrated by experiments involving polarization analysis of photofragment fluorescence from the groups of Paul L. Houston at Cornell University, Stephen R.

Crossed-beam experiments reveal reaction dynamics

The state-of-the-art of both theoretical and experimental chemical reaction dynamics can be seen in recent work based on crossed-beam methods on two elementary reactions. The best-known example is the hydrogen exchange reaction, $H + H_2 \rightarrow H_2 + H$ (below). The potential energy surface for this reaction and its various isotopic counterparts was computed, point by point, by Per E. M. Siegbahn of the University of Stockholm, Sweden, and Bowen Liu of the IBM research laboratory in San Jose, Calif., using *ab initio* quantum chemical procedures believed to give energies with an accuracy of 0.1 kcal per mole. These data were later fitted to analytical functions by Donald G. Truhlar and Charles J. Horowitz at the University of Minnesota to make them easier to use. This "best-known" potential energy surface is designated LSTH, and extensive quantal computational advances using it have been made by the groups of Aron Kuppermann at California Institute of Technology, Robert E. Wyatt at the University of Texas, Austin, and John C. Light at the University of Chicago. The lower panel shows calculations for the expected angular distribution of the product molecule HD from the $H + D_2$ reaction, using quasiclassical trajectory calculations on the LSTH surface.

Such calculations by Truhlar and Normand C. Blais of Los Alamos National Laboratory and by Howard R. Mayne at the University of New Hampshire predict very well the experimental detailed differential reactive scattering cross-sections for this same reaction obtained experimentally by J. Peter Toennies and coworkers at Max Planck Institute for Fluid Dynamics, Göttingen, West Germany, shown in the upper panel. In an important 1986 paper, they showed that the HD angular distributions are strongly anisotropic (in the scattering angle θ), implying a short-lived transition state. Quasiclassical trajectory calculations of vibrational and rotational state distributions of the HD product are also in reasonable accord with laser experiments by the groups of Richard N. Zare at Stanford University and James J. Valentini at the University of California, Irvine.

The reactive cross sections shown here are for a collision energy of 1.5 eV. These experimental results are not only mimicked by the quasiclassical trajectory calculations, but by a simple classical impulsive model consistent with a transition state that lasts for a very short time. Very recently, Valentini has observed sharp structure in the energy dependence of the reaction cross-section for specific vibrational states of the products. Such dynamical resonances have

Experimental reactive-scattering contour diagram in the center of mass representation for the $D + H_2$ crossed-beam reaction. The HD product flux–velocity is seen to be ejected preferentially sideways with respect to the relative velocity (of about 3 km per second). [θ is the product-scattering angle in the center of mass system, u is the velocity in the center of mass system, and v is the velocity in the laboratory system.] The results are well represented by theory (lower panel). [Adapted from R. Götting, H. R. Mayne, and J. P. Toennies, *J. Chem. Phys.*, **85**, 6396 (1986).]

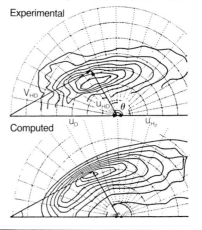

Experimental

Computed

Experimental product flux–velocity contour map in the center of mass system for DF from the F + D_2 reaction at two different collision energies. The circles denote velocity limits for the specified DF vibrational states. For the lower collision energy case, DF is predominantly backward scattered, that is, antiparallel to the incident F velocity, for all observed vibrational states. For the higher collision energy case, the DF is strongly forward scattered (attributed to resonance) for the $v = 4$ state. [Adapted from D. M. Neumark, A. M. Wodtke, G. N. Robinson, C. C. Hayden, K. Shobatake, R. K. Sparks, T. P. Schafer, and Y. T. Lee, *J. Chem. Phys.*, **82**, 3067 (1985).]

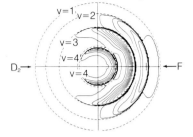

Collision energy = 1.82 kcal per mole

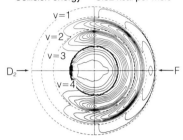

Collision energy = 3.32 kcal per mole

been confirmed theoretically by Truhlar and Donald J. Kouri of the University of Houston.

The second important bimolecular reaction that has been studied is F + H_2 → HF + H and its isotopic variants. A much larger experimental chemical dynamical database is available for this reaction, the most detailed of which comes from the crossed molecular beam studies of Yuan T. Lee and coworkers at the University of California, Berkeley (above). The upper panel shows a typical flux–velocity contour map plotted in the center-of-mass system, indicating backward scattering of the DF product in vibrational states 1, 2, 3, and 4. But at a different collision energy (lower panel) there is a drastic change in the angular distribution, which now shows a forward peak for the $v = 4$ state. This is a clue that quasibound states are formed in the encounter, which decay preferentially into product molecules in this vibrational energy state. (Analogous experiments with F + H_2 show such an anomaly of another energy, with a forward HF peak for $v = 3$.)

Various semiempirical and *ab initio* potential energy surfaces have been put forward by James T. Muckerman of

Brookhaven National Laboratory, by John C. Polanyi at the University of Toronto, and by Henry F. Schaefer III at UC Berkeley, which can account for many of the gross dynamical features of the reaction and even predict certain resonances in the scattering cross-sections. However, as yet no calculation does full justice to the experimental, vibrationally specific product angular distributions measured over a range of collision energies for the three isotopic reactions of F with H_2, HD, and D_2. Very recently, William H. Miller and coworkers at UC Berkeley have obtained accurate quantal scattering computational results, which perhaps can be extended to deal with the entire body of experimental observations. Because of the mainly direct characteristics of the reaction, it is clearly one in which the lifetime of the transition state is in the femtosecond rather than the picosecond regime, except for the long-lived resonances. As Lee points out, one of the as yet incomplete tasks of the molecular beam method is the "direct experimental probing of potential energy surfaces."

Leone at the University of Colorado, John P. Simons at the University of Nottingham, U.K., and Richard N. Dixon at the University of Bristol, U.K.

In the case of half-collisions, subpicosecond and femtosecond measurements of the dissociation lifetime have been obtained by the Caltech group for several elementary, laser-initiated, unimolecular reactions, and these temporal results have been compared with those from spatial (orientation) experiments.

In certain polyatomic molecules, such as the methyl halides, a quasidiatomic approximation can account for their continuous absorption spectra. However, to interpret the photofragment distribution data, it is necessary to take into account the internal excitation of the fragments. Various simple classical models for half-collisions have been useful in accounting for experimental data on fragment energy distributions, but classical trajectory simulations provide more insight, just as they do for full collisions. The most popular approach for the description of dissociation trajectories has been that of simple impulsive forces between fragments. One imagines instantaneously breaking the bond and then makes classical mechanical predictions of what would follow, using conservation of energy and momenta. For example, if a torque is produced, then one would expect rotational excitation of the fragments.

Even better, of course, would be exact quantal calculations on *ab initio* potential surfaces. As is also the case for full collisions, these calculations are just a little beyond the present state of the art. An extra difficulty with photofragmentation reactions is the need to calculate not just the ground-state potential energy surface, but also that of the excited state and any interactions with other, possibly nearby or crossing, surfaces. Even in the simplest case, where there is only one excited state, there are problems in the implementation of time-independent quantum mechanical procedures. However, experimental advances by many research groups in the U.S., West Germany, France, Switzerland, and the U.K. have aided the application of theory, particularly to experiments on the photodissociation reactions of simple molecules, such as HOH, HOOH, HONO, NCNO, ICN, CH_3I, and CH_3ONO.

Product state distributions are expected to be sensitive mainly to the nature of the potential surface in the region of product separation. On the other hand, the rates of these reactions represent the flux of excited molecules traversing the transition state and therefore relate to the nature of the surface near the critical transition state configuration. Real-time picosecond measurements of state-to-state rates have now been made and can be used to test theories of unimolecular reactions and to help elucidate the nature of the potential energy surface near the transition state.

Experimental state-to-state (microcanonical) rate constants for the reaction $NCNO \rightarrow NC + NO$ have been measured at Caltech as a function of energy above threshold for bond rupture. For this same reaction, Curt Wittig and Hanna Reisler of the University of Southern California have made a thorough study of the NC and NO product state distributions. The microcanonical rate constants show an unexpected nonmonotonic structure in the energy dependence, and deviate significantly from theoretical calculations based on the phase space theory that was used successfully to calculate product state distributions. Similar studies of the unimolecular reaction $HOOH \rightarrow 2\ OH$ also have been carried out by F. Fleming Crim of the University of Wisconsin (product state distribution) and the Caltech group (real-time rates).

The phase space theory, the adiabatic channel model, and the RRKM theory for describing unimolecular reactions can now be critically tested on the microcanonical, molecular level. A particularly good example is recent work on the photofragmentation of ketene by the group of C. Bradley Moore at the University of California, Berkeley, who examined product state distribution near threshold energy, and the group at Caltech who determined the state-to-state rates for the same reaction. They find that although phase space theory agrees with observations near threshold, it must be modified to account fully for the dynamics at all energies. This problem is currently being addressed theoretically by the groups of Rudolph A. Marcus of Caltech, William H. Miller at UC Berkeley, Jürgen Troe at the University of Göttingen, West Germany, and others. Progress will be achieved when the topography of the potential energy surface involved is better known from *ab initio* quantum mechanical computations and when the quantum dynamics can be solved.

Although these studies have provided many details of the dynamics, neither the state-to-state rate measurements nor the product state distribution data provide a direct view of the transition states.

Photofragment spectroscopy experiments reveal reaction dynamics

The photofragment spectroscopy of three- and four-atom molecules presented here shows how dynamical information is extracted and is typical of many studies, made in the U.K., Germany, Switzerland, and the U.S.

The group of C. Bradley Moore at the University of California, Berkeley, has provided a large body of experimental data on the photodissociation of formaldehyde that has helped to establish how the excited formaldehyde molecule breaks apart to form molecular hydrogen and carbon monoxide (and also atomic hydrogen plus the formyl radical). The path for this photodissociation is shown below. The initial, excited state of formaldehyde is well characterized by spectroscopic studies. Internal conver-

sion causes coupling between the first excited singlet state of the molecule (S_1) and the quasidegenerate rovibrational levels of the ground state ($S_0{}^{**}$). Consequently, the initial state of the reaction cannot be considered to be a discrete state, but rather a lumpy continuum of states. Molecules in these states reach the reaction coordinate by a process called intramolecular vibrational-energy redistribution, and such processes have been characterized experimentally and theoretically for many molecules.

Experiments have revealed the entire rotational and vibrational distribution of both the carbon monoxide and hydrogen products. By energy conservation, the remaining rather large fraction of the available energy is known to go into the

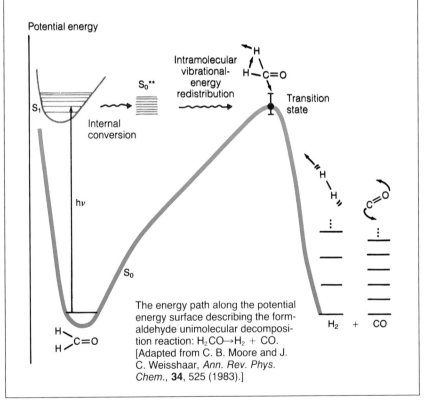

The energy path along the potential energy surface describing the formaldehyde unimolecular decomposition reaction: $H_2CO \rightarrow H_2 + CO$. [Adapted from C. B. Moore and J. C. Weisshaar, *Ann. Rev. Phys. Chem.*, **34**, 525 (1983).]

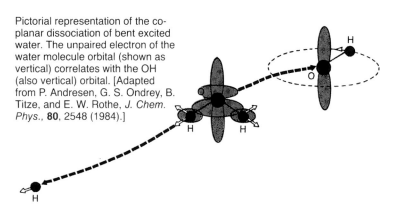

Pictorial representation of the coplanar dissociation of bent excited water. The unpaired electron of the water molecule orbital (shown as vertical) correlates with the OH (also vertical) orbital. [Adapted from P. Andresen, G. S. Ondrey, B. Titze, and E. W. Rothe, *J. Chem. Phys.*, **80**, 2548 (1984).]

relative translation of energy of the product molecules, as verified by molecular beam translational spectroscopy experiments. The nature of the transition states in this reaction has been deduced by Henry F. Schaefer III and coworkers, then at UC Berkeley, from computations of relevant portions of the potential energy surface. Dynamical calculations utilizing this surface by William H. Miller and coworkers, also at Berkeley, reproduce the main features of the experimental energy disposal patterns. These and other studies show how state-to-state population analysis can help establish a microscopic mechanism for the photofragmentation. Theoretical work by Karl F. Freed at the University of Chicago, Moshe Shapiro at Weizmann Institute of Science in Israel, Richard Bersohn at Columbia University, David A. Micha at the University of Florida, Keiji Morokuma at the Institute for Molecular Science in Japan, and others also contributed toward the understanding of photofragmentation.

The energy path for the photodissociation of a second polyatomic molecule, water, shown above, is much different. This photofragmentation has been thoroughly studied experimentally by Peter Andresen of Max Planck Institute for Fluid Dynamics in Göttingen, West Germany and John P. Simons at the University of Nottingham, U.K. The angular distribution, population, and rotational alignment of the fragments produced after excitation to the absorption continua have been characterized with the help of theoretical developments by Reinhard Schinke of Max Planck Institute and Richard N. Dixon at the University of Bristol in England. A great deal now is known about the way this molecule, which is bent in its excited state, dissociates.

The upper excited potential energy surface for water is repulsive. The photoinduced reaction, at 157 nm, can be described as

$$H_2O \xrightarrow{h\nu} H_2O^* \to H + OH.$$

The OH photofragment molecules are vibrationally hot but rotationally only "warm." A considerable fraction of the available energy in the reaction goes into translational recoil between the H and OH fragments. When the water molecule bond breaks, there are two possibilities for the rotation of OH: either in the same plane as the oxygen p-orbital (π^- state) or perpendicular to it (π^+ state). Experiments show that the dissociation leads to preferential population of π^+. This OH alignment leads to the conclusion that the photofragmentation process is essentially coplanar, as depicted in the figure. This inverted population for the lambda doublet explains the origin of interstellar OH maser emission formed via ultraviolet irradiation of water molecules in the galactic environment.

UNI- VERSUS BIMOLECULAR REACTIONS

Unimolecular and bimolecular reactions have many dynamic aspects in common, and the connection between these two classes of reactions is important to femtochemistry. A few examples illustrate this. Hershbach showed how the product angular distribution in the reaction $Cl_2 + H \rightarrow HCl + Cl$ reveals the dependence of the reaction probability on the relative orientation of colliding partners in the transition

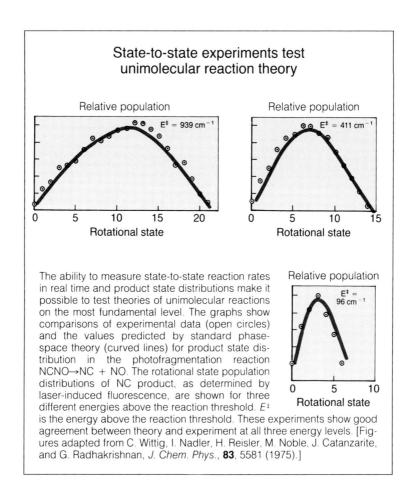

State-to-state experiments test unimolecular reaction theory

The ability to measure state-to-state reaction rates in real time and product state distributions make it possible to test theories of unimolecular reactions on the most fundamental level. The graphs show comparisons of experimental data (open circles) and the values predicted by standard phase-space theory (curved lines) for product state distribution in the photofragmentation reaction NCNO→NC + NO. The rotational state population distributions of NC product, as determined by laser-induced fluorescence, are shown for three different energies above the reaction threshold. $E^{\ddagger}$ is the energy above the reaction threshold. These experiments show good agreement between theory and experiment at all three energy levels. [Figures adapted from C. Wittig, I. Nadler, H. Reisler, M. Noble, J. Catanzarite, and G. Radhakrishnan, *J. Chem. Phys.*, **83**, 5581 (1975).]

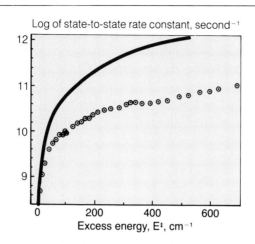

Log of state-to-state rate constant, second⁻¹

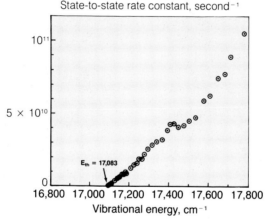

State-to-state rate constant, second⁻¹

Above are the results of experiments measuring the state-to-state rate constant for the same reaction. The top graph is a comparison of data from a picosecond laser pump-probe experiment with values predicted by standard phase-space theory. Note that the real-time data are not well represented by theory in this experiment, particularly at high energies. The bottom graph shows the experimental values for the rate constant as a function of vibrational energy in the reagent. These points clearly do not fit a smooth curve. Product state distribution studies and state-to-state rate studies examine different portions of the reaction potential energy surface. [Figures adapted from L. R. Khundkar, J. L. Knee, and A. H. Zewail, *J. Chem. Phys.*, **87**, 77 (1987).]

state. He also showed the close analogy between the dynamics of this type of direct-mode, atom–molecule reaction and the corresponding photon–molecule dissociation reaction: The HCl product recoil velocity distribution from the H + Cl$_2$ reaction is much like that of the product chlorine atoms in photodissociation. Because the photofragment angular distribution mimics the orientation of the photo-excited HCl molecule (or that of the transition state HClCl) before it rotates, the dissociation is described as prompt; that is, it occurs before the molecule (or transition state) has time to rotate.

This example indicates the connection between an ultrafast photon–molecule reaction and the analogous atom–molecule reaction in which

Product angular distributions give clues to transition state lifetimes

H + Cl$_2$ ⟶ HCl + Cl

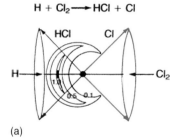

(a)

hν + Cl$_2$ ⟶ Cl + Cl

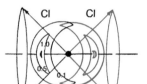

(b)

Studies of product distribution in unimolecular and bimolecular reactions reveal reaction dynamics. Figures (a) and (b) compare the product flux–velocity distributions for a bimolecular reaction, H + Cl$_2$→HCl + Cl (a) and the unimolecular photofragmentation reaction, hν + Cl$_2$→Cl + Cl (b). For the bimolecular reaction, the H reactant approaches from the left, the Cl$_2$ from the right. After collision, most of the product HCl molecules recoil to the left; the red cone encloses the region of maximum HCl intensity. Similarly, the Cl product recoils to the right into the region enclosed by the blue cone. Superimposed is a contour map indicating the angular and recoil velocity distributions for the HCl product. (Higher numbers indicate greater concentration of product molecules.) For the photodissociation of Cl$_2$, the cones and contour maps show a very similar distribution for the two Cl atom products.

the transition state has a subpicosecond lifetime. A more general relationship between unimolecular and bimolecular reactions is illustrated by two examples involving a long-lived complex: the Hg + $I_2 \rightarrow$ [IHgI] $\rightarrow$ HgI + I system studied by the group of Richard B. Bernstein, now at the University of California, Los Angeles, using the crossed-molecular-beam technique, and the K + NaCl $\rightarrow$ [KClNa] $\rightarrow$ KCl + Na system studied by the group of Philip R. Brooks and Robert F. Curl of Rice University, for which the collision complex was detected by laser-induced fluorescence.

For the system K + NaCl, a laser excites the bimolecular-formed KClNa collision complex. Independent evidence from the study of clus-

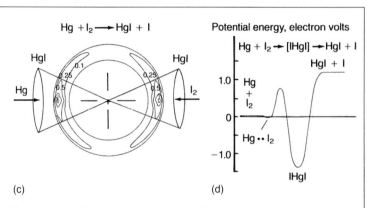

(c)

(d)

These patterns (figures (a) and (b)) are quite different from that for the reaction of Hg + $I_2 \rightarrow$HgI + I, shown in figure (c). Here HgI product recoils both forward and backward with respect to the center-of-mass (at the origin of the contour map). This forward–backward symmetry is characteristic of a long-lived collision complex that undergoes many rotations before it breaks apart. For the reactions in figures (a) and (b), by contrast, the collision complex lasts such a short time that it does not have time to rotate before it breaks apart. Figure

(d) is the experimentally derived potential energy curve for the Hg + I_2 reaction. Note the deep well corresponding to the binding energy of the IHgI collision complex, another indication that the complex will be long-lived. [Figures (a) and (b) adapted from D. R. Herschbach, *Far. Disc. Chem. Soc.*, **55**, 241 (1973). Figures (c) and (d) from M. M. Oprysko, F. J. Aoiz, M. A. McMahan and R. B. Bernstein, *J. Chem. Phys.*, **78**, 3816 (1983) and T. M. Mayer, J. T. Muckerman, B. E. Wilcomb, and R. B. Bernstein, *J. Chem. Phys.*, **67**, 3522 (1977).]

ter beams of mixed alkali halides by the group of Ernst Schumacher at
the Institute for Inorganic and Physical Chemistry in Berne, Switzerland, has shown that a relatively stable KClNa complex can be formed
in a jet-cooled beam. These are good candidates for future femtochemistry experiments to establish connections between unimolecular and
bimolecular reactions on the same potential energy surface.

TIME-INTEGRATED SPECTRA OF THE TRANSITION REGION

In recent years, there has been great progress in using time-
integrated spectroscopic techniques to probe reactions during the reactive process. Absorption, emission, scattering, and ion spectroscopy
methods have all been utilized. The spectra of transient species intermediate between reagents and products are expected, of course, to be
different from those of the stable products or reagents. For one thing,
transition energies and spectral linewidths for these intermediate species will be shifted and broadened because of perturbations caused by
the proximity of product fragments, as shown for many systems including those studied by Alan Gallagher and coworkers, then at the University of Colorado, the group of William C. Stwalley at the University
of Iowa, and others.

John C. Polanyi and his collaborators have used the emission in the
course of a unimolecular reaction

$$\text{NaI} \xrightarrow{\text{uv}} \text{NaI}^{\neq *} \to \text{Na}^* + \text{I}$$

and of a bimolecular reaction

$$\text{F} + \text{Na}_2 \to [\text{NaNaF}]^{\neq *} \to \text{Na}^* + \text{NaF}$$

to characterize their transition-state regions. Very recently, femtochemistry observations also have been made at Caltech on the first of
these reactions. For the bimolecular reaction, no lasers need be used
since the sodium product Na^* emits light. The transition state species
$[\text{NaNaF}]^{\neq *}$ has a lifetime of one picosecond or less, while the lifetime
of Na^* is about 10^{-8} second. Therefore, the light emitted from the transition state is about 10^{-4} the amount emitted from the Na product.

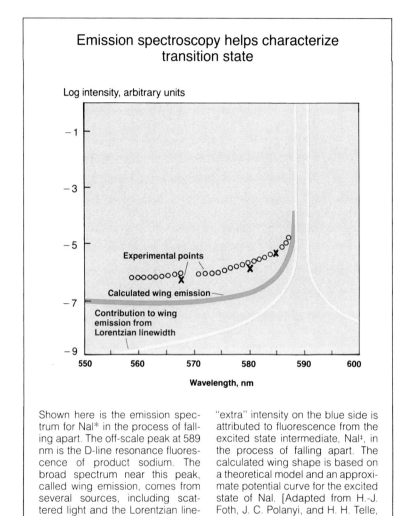

Emission spectroscopy helps characterize transition state

Shown here is the emission spectrum for NaI* in the process of falling apart. The off-scale peak at 589 nm is the D-line resonance fluorescence of product sodium. The broad spectrum near this peak, called wing emission, comes from several sources, including scattered light and the Lorentzian linewidth of the D-line. The observed "extra" intensity on the blue side is attributed to fluorescence from the excited state intermediate, NaI‡, in the process of falling apart. The calculated wing shape is based on a theoretical model and an approximate potential curve for the excited state of NaI. [Adapted from H.-J. Foth, J. C. Polanyi, and H. H. Telle, *J. Phys. Chem.*, **86**, 5027 (1982).]

Spectral data show wing emission that extends over several hundred angstroms from the sodium D line. Thus, the relative intensity of the two emissions at a given wavelength is less than the expected 10^{-4} by still another factor of 100 or so because of the spectral breadth of the wings. Polanyi's group has deduced some potential models that can account for these observations. In a 1987 study, Polanyi examined the hot atom reaction of $D + H_2$, and found evidence for a $DH_2^{\ddagger}$ transient species. Such experiments are clearly relevant to the important H_3 family of reactions.

In 1984, James L. Kinsey and coworkers, then at Massachusetts Institute of Technology, showed that measurements of near-resonant Raman scattering spectra from laser-excited methyl iodide or ozone molecules in the very process of dissociation can be used to characterize the dynamics along the reaction coordinate. The concept involved is best visualized with the help of wave packet theory, developed by Eric J. Heller, then at UCLA. Basically, if a laser prepares a packet of excited molecules on an upper potential energy surface, this packet will in time spread out and move to larger internuclear distances, leading to photodissociation. The Raman-shifted spectral wavelengths will depend on the ground-state characteristics, but the intensities of these spectral transitions will reflect the evolution of the wave packet on the upper surface. From these spectra, Kinsey and coworkers deduced the nature of the excited-state surface.

Brooks and Curl have invoked a different method to study the transient intermediate in bimolecular reactions. As with the two previous examples, the lasers used have pulse durations of several nanoseconds or are in a continuous-wave mode. The collision complex, however, lasts on the order of picoseconds. Thus, light absorption during its lifetime will be extremely small. Such experiments have been carried out on the reaction

$$K + NaCl \xrightarrow{h\nu} [KClNa]^* \rightarrow KCl + Na.$$

The laser in this case was tuned to wavelengths where neither reagents nor products absorb, so the signal is attributed to the transient complex. As Brooks pointed out recently, "The problem facing the experimentalist [when the yield of emission or absorption is very small] is to design a system where the effect of photon absorption [or emission] during collision can be observed and separated from other more 'mundane' effects."

Despite these difficulties, great progress has been made. The spectra have provided valuable experimental insight about the transition

Wave packet propagation describes the process of photofragmentation

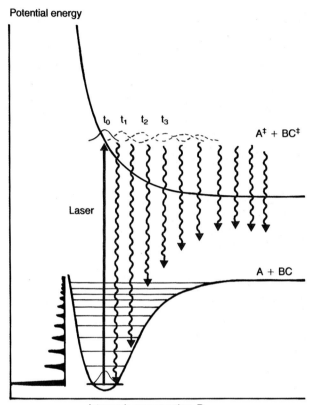

Potential energy

t_0 t_1 t_2 t_3

$A^{\ddagger} + BC^{\ddagger}$

Laser

A + BC

Internuclear separation, R_{AB}

Simplified potential energy curves describe the photodissociation of a hypothetical molecule, ABC, into A + BC. The lower curve shows the ground state and the upper one the excited, dissociative state. In the description provided by wave packet theory, the laser photon transfers the ground state wave function to the excited state "packet" at the time shown as t_0.

During the process of falling apart, an excited ABC molecule can fluoresce to various vibrational states of the ground state. This emission spectrum, shown in the lower left corner, consists of a series of broadened lines spaced according to the ground state vibrational levels. [Adapted from D. Imre, J. L. Kinsey, A. Sinha, and J. Krenos, *J. Phys. Chem.*, **88**, 3956 (1984).]

Laser excites transient [KClNa] complex

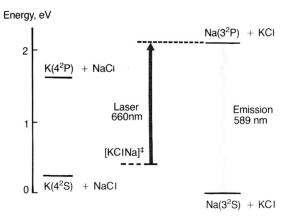

The concept of an experiment to observe the light absorption by the collision complex in the reaction K + NaCl→Na + KCl is illustrated on this energy-level diagram. 660-nm laser photons excite the [KClNa]‡ collision complex. Since the photon energy is not "in resonance," it cannot excite either the reagents or the products. Thus the absorption, manifested by the Na D-line emission at 589 nm and observed only when the two reagent beams cross and the laser beam intersects the scattering zone, is due to the long-lived [KClNa]‡ collision complex. [Adapted from T. C. Maguire, P. R. Brooks, R. F. Curl, J. H. Spence, and S. Ulvick, *J. Chem. Phys.*, **85**, 844 (1986).]

region, beyond that obtained from more conventional methods. Implicit in these measurements, but unresolved, are the femtosecond dynamics.

REAL-TIME FEMTOCHEMISTRY

Since transition states exist only for picoseconds or less, catching them in real time requires femtosecond time resolution. As already discussed, in the steady state, the amount of light emitted from transition-state molecules is smaller by four to six orders of magnitude than that from the emission of free fragments. Thus, if ultrashort pulses are employed, the sensitivity of the detection is enhanced by a

Real-time femtochemistry experiment catches excited ICN molecule in act of falling apart

Potential energy, cm^{-1} × 10^4

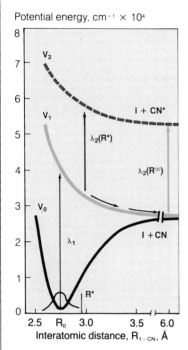

Interatomic distance, R_{1-CN}, Å

Above: The drawing shows the three relevant potential energy curves of ICN simplified to describe a femtochemistry study of its photodissociation. The lowest curve, V_0, refers to the ground state molecule, the middle one, V_1, to the lowest-lying repulsive excited state. The pump laser photon (λ_1) excites the ICN from V_0 to V_1 at R_0. Here, R_0 is the equilibrium bond length I—CN. The probe laser pulse, whose wavelength is set for the free CN resonance, $\lambda_2(R^\infty)$, is delayed by variable time intervals with respect to the pump pulse. The laser-induced fluorescence intensity remains zero until the CN has separated to about 6 Å from the I atom. The probe laser pulse, now de-tuned to the red to $\lambda_2(R^*)$, and delayed by a time interval measured in femtoseconds, excites the ICN at $V_1(R^*)$, as the molecule is in the process of falling apart (when the separation between I and CN has reached the distance R^*), to a higher electronic state, V_2, which fluoresces. Thus the intensity of the fluorescence increases with time to a peak and then quickly falls as the distance between the I and CN fragments increases and the probe laser wavelength is no longer in resonance.

Below: Pump-probe transients are observed by laser-induced fluorescence of the CN photofragment. Upper: When the probe laser is tuned to the wavelength ($\lambda(R^*)$) corresponding to free CN, the signal is delayed until the I and CN have "fully" separated. Lower: The probe laser is detuned to the red to detect the perturbed CN in the process of separation from the I atom. The clocking of this reaction from $t = 0$ to complete separation of I and CN yields a 205 ± 30 fs delay time. [Adapted from M. Dantus, M. Rosker and A. H. Zewail, *J. Chem. Phys.*, **87**, 2395 (1987) and M. Rosker, M. Dantus, and A. H. Zewail, *Science*, **241**, 1200 (1988).]

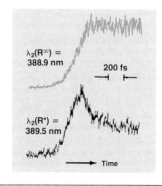

$\lambda_2(R^\infty) = 388.9$ nm

200 fs

$\lambda_2(R^*) = 389.5$ nm

Time

factor of 10^4 to 10^6 when compared with time-integrated experiments. Essentially all molecules traversing the transition states can be detected; this is a key feature of femtochemistry. The difficulty, however, comes in generating pulses and in developing techniques for observing changes on the femtosecond time scale.

Real-time femtochemistry started with the subpicosecond experiments on the photodissociation of ICN reported in 1985 from Caltech. The time resolution was then improved and the technique was extended to detect transition states by changing the probe wavelength to the absorption of perturbed fragments in transit. The resulting spectra show the CN fragment in the process of separation from the iodine atom in the reaction

$$ICN^* \rightarrow [I \cdots CN]^{**} \rightarrow I + CN.$$

The same technique has recently been applied at Caltech to reactions of alkali halides, where resonance trapping of the departing atoms was observed

$$NaX^* \rightarrow [Na \cdots X]^{**} \rightarrow Na + X \qquad (X = I, Br).$$

It is interesting that this first experimental observation of trapping resonances in dissociation reactions, which helped establish the "femto age," dealt with systems from the "alkali age" of crossed-molecular-beam studies.

The idea is as follows: The pump pulse excites a target molecule, say ABC, to a dissociative state. The probe pulse, delayed by a variable time, detects the photofragment product, say AB, as it is being formed, in the process of separation from C. The probe laser is first tuned to a wavelength corresponding to a known excitation resonance of the stable AB species, either yielding laser-induced fluorescence or bringing about multiphoton ionization. In either case, the excitation allows detection of AB. The resulting photon or photoion signal is recorded as one point on a curve of intensity (measuring concentration) as a function of time. The delay is altered systematically until an entire curve for this wavelength is obtained. There is an induction period corresponding to the time required for the AB species to separate effectively from the force field of C and to attain, asymptotically, its normal identity as AB in its final rotational and vibrational energy states.

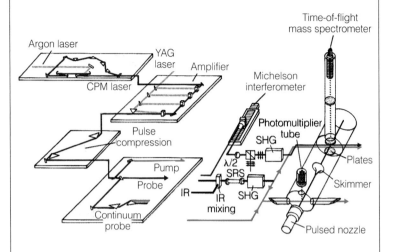

Femtosecond lasers combine with molecular beams for femtochemistry experiments

Caltech apparatus for femtochemistry and picochemistry experiments includes, at left, colliding-pulse-mode (CPM) locked ring laser, YAG-pumped laser amplifier system, pulse compression system, and an arrangement for pump and probe pulse generation. At right is the molecular beam system, including the time-of-flight mass spectrometer, interfaced to the laser system. The delay of the probe pulse is set by the Michelson interferometer arrangement. Shown also are the different nonlinear techniques—second harmonic generation (SHG), stimulated Raman scattering (SRS), and infrared mixing—used in general in these pump and probe femtochemistry experiments. The polarization of the pump and probe laser beams can be adjusted separately. For some experiments a flowing gas system is used instead of the molecular beam. [Adapted from M. Rosker, M. Dantus, and A. H. Zewail, *J. Chem. Phys.*, **89**, 6113 (1988).]

The delay curve is repeated, detuning the probe laser by small successive increments to reach the fragment absorption wavelength in the transition region. At each wavelength, a buildup and decay of the transition state is recorded. A typical curve goes through the maximum at short delay times and then decays asymptotically to a constant signal level, dependent upon the detuning increment. This entire set of experiments can be repeated at different pump pulse wavelengths, thus changing the available energy for product recoil. The resulting data constitute a surface from which can be deduced the potential energy for the formation of AB from the excited state of ABC.

Two technologies are involved in much of femtochemistry (and the earlier, sister picochemistry) studies under collisionless conditions: molecular beams and the generation and characterization of ultrashort laser pulses. The laser system built at Caltech is based on the pioneering design provided in 1981 by Shank and colleagues at AT&T Bell Laboratories. They have shown that 90-fs pulse widths at 100-MHz repetition rate can be obtained from a colliding-pulse-mode locked ring dye laser.

With the help of pulse compression and group velocity compensation techniques, a number of groups have generated shorter femtosecond pulses; the shortest is 6 fs by the AT&T group. These pulses can be amplified, and pulse broadening caused by group velocity dispersion in the amplifier can be eliminated by a special optical arrangement.

Part of Caltech's femtosecond apparatus, including its laser components, is mounted on a floating optical table that measures 20 feet × 5 feet.

Thus, femtosecond pulses are generated at different wavelengths, opening up a number of possibilities for studies on this time scale. The characterization of the pulses is made by standard auto- and cross-correlation techniques. For subpicosecond and picosecond experiments, two synchronously pumped dye lasers (with pulse compression capability) are used in similar arrangements.

For experiments on rotationally and vibrationally cold molecules, a standard molecular beam source is used (employing a pulsed, seeded supersonic beam) with electron impact and laser ionization, time-of-flight mass spectrometer, and laser-induced fluorescence capabilities. As shown by Zare, laser-induced fluorescence is a very sensitive method of detection, and the work of Donald H. Levy, Lennard Wharton, and Richard E. Smalley at the University of Chicago showed that cooling of molecules simplifies the spectra greatly. Laser ionization time-of-flight mass spectroscopy, as introduced by Bernstein, then at Columbia University, and Edward Schlag at the Technical University in Munich, helps identify the fragment of interest. Thus, mass and quantum state resolution is possible. The laser beams intersect the molecular beam

(or the gas) in a very small interaction zone. The time at which the initial laser pulse arrives at this zone marks the beginning of the experiment and establishes the zero of time; the probe pulse follows.

The old concept of Michelson interferometry is used to clock the experiments. Time is determined by controlling the distance traveled by the two pulses (3 μm distance is equivalent to 10 fs), starting when the two pulses overlap. For each experiment at a chosen delay time, typically in the range of -100 to $+1000$ fs, the detected ionization or fluorescence signal is integrated for a sufficient length of time to yield a measure of total intensity. The experiments are repeated at different delays and the so-called "transient" (a curve of intensity as a function of time) is constructed.

In these experiments, probe pulse wavelengths are generated by nonlinear methods, using doubling and mixing crystals, or by producing a continuum of different wavelengths by focusing the femtosecond pulse on water in a jet or a cell. The same can be done for the pump laser, and the desired wavelengths are selected by a monochromator or by using interference filters. The entire experiment is repeated at each new wavelength of interest to construct the spectra of the reaction at a given available energy and at different probing wavelengths. The Caltech machines utilize 40 fs pulses from the colliding pulse-mode laser, 60 to 100 fs pulses from the amplified colliding pulse-mode laser system, and subpicosecond or 2- to 5-picosecond pulses from the synchronously pumped dye laser systems.

FEMTOCHEMISTRY OF UNIMOLECULAR REACTIONS

As mentioned previously, the first of these femtosecond studies dealt with unimolecular reactions. The experiments were done on the elementary reaction of ICN $\rightarrow$ I + CN. The pump pulse was at 307 nm and the probe was set at a wavelength of 388.5 nm (the absorption peak for free CN fragments) or detuned by as much as 10 nm to detect perturbed CN at these wavelengths, via its laser-induced fluorescence. The signal is proportional to the number of perturbed CNs. As anticipated, when tuning to the perturbed CN fragment absorption, the transients exhibit a buildup and a decay characteristic of the short-lived (about 10^{-13} second) transition states. On-resonance absorption of the free CN fragment gives the time for dissociation of the ICN into the fragments, 205 ± 30 fs.

The time for bond breaking depends on the characteristics of the potential energy surface, and the observed transients provide a way to view the dynamics on these potential energy surfaces. Basically, the femtosecond probe pulse "sees" the potential at different intramolecular separations.

For a simple, classical bond-breaking process, assuming pure exponential repulsive forces, the change of the potential with time and then the corresponding change with intermolecular separation is obtained directly from the data. Such classical modeling by Bersohn and Zewail, as well as a quantum-wave-packet treatment of the dissociation by Dan G. Imre at the University of Washington, can reproduce the main features of the experimental observations and provide potential-energy-surface parameters. In general, however, the potential energy surfaces are more complex and the data must be inverted to obtain the shape. We have recently developed such an inversion procedure.

Progress in femtochemistry does not require new developments in laser technology. For transform-limited (Gaussian) pulses at the wavelength of the ICN experiments, the spread of the wavelength–intensity distribution of the pump laser is 3.5 nm for a 40-fs pulse; for the probe laser, 4.4 nm. If spectroscopic resolution is required, no advantage is to be gained from further shortening the laser pulse. In fact, because of the uncertainty principle, it is of little value from a chemical viewpoint to use pulses much shorter than those now available. For example, at $\lambda = 308$ nm, 7 fs corresponds to 20 nm; thus the energy of the excitation is $E = 4.02$ eV, with an uncertainty of ± 0.13 eV. This is a total energy width of 0.26 eV, or 6 kcal per mole. It is soon going to be necessary to compromise between temporal and energy resolution. From a chemical dynamics point of view, since intermolecular and intramolecular atomic velocities are usually of the order of magnitude of 1 km per second, a time interval of 1 fs corresponds to an atomic displacement of about 0.01 Å. This is certainly "slow motion." Even for the fastest reaction of all, $H + H_2$, the necessary time resolution is only about 10 fs.

FEMTOCHEMISTRY OF BIMOLECULAR REACTIONS

These examples illustrate the potential to conduct real-time femto-chemistry observations of the dynamics of a wide variety of photon-induced unimolecular processes, both dissociations and isomerizations. But the field of chemical dynamics is broader than this. The central theme of chemistry is the transformation of a set of reactant molecules into product molecules. The overall mechanism of reactions usually involves a number of elementary steps, many of which are bimolecular in nature. In the $H_2 + F_2$ chemical laser reaction, for example, the key elementary step is the fast abstraction of a hydrogen atom from the hydrogen molecule: $F + H_2 \rightarrow HF + H$. In this bimolecular reaction, an H—H bond is transformed into an H—F bond via a three-center transition state in such a way that the nascent HF product is vibrationally and rotationally excited.

Molecular beam studies of reactive scattering have characterized a great many such elementary bimolecular reactions in terms of post-collision attributes, such as energy distribution in the products and linear and angular momentum distribution. These attributes represent asymptotic properties that have been time-integrated over the course of the collision. It is highly desirable to be able to observe the collision itself in real time.

An important new issue arises for bimolecular reactions that is not a concern in the unimolecular case. For a unimolecular reaction, the reaction begins when the laser pulse initiates excitation and subsequent dissociation. Thus, the interval between the firing of this laser pulse and the subsequent probe laser pulse represents the actual amount of time that the reaction has been taking place. But for a bimolecular reaction, whether in bulk or in crossed beams, there is no comparable way to establish the beginning point for the reaction. Even if one attempted to initiate a reaction by using a short laser pulse to form radicals that can then react (such as, for instance, in the reactions $R + R'H \rightarrow RH + R'$, or $R' + R \rightarrow RR'$), the radicals would first have to find each other and collide before reaction can possibly occur.

In gas-phase systems, the average time between collisions is inversely proportional to the pressure; at subatmospheric pressures, these times are typically in the nanosecond to microsecond range,

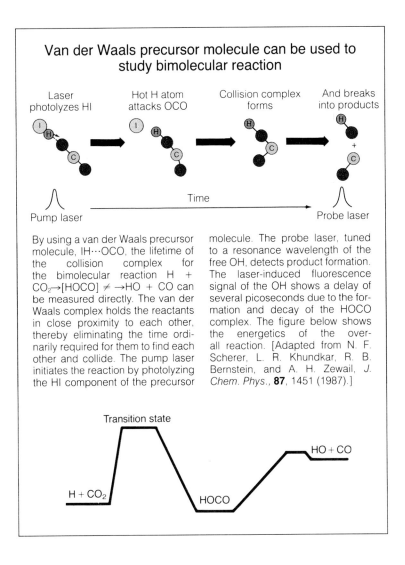

Van der Waals precursor molecule can be used to study bimolecular reaction

Laser photolyzes HI | Hot H atom attacks OCO | Collision complex forms | And breaks into products

Pump laser — Time → Probe laser

By using a van der Waals precursor molecule, IH···OCO, the lifetime of the collision complex for the bimolecular reaction H + $CO_2 \rightarrow$ [HOCO] $\neq \rightarrow$ HO + CO can be measured directly. The van der Waals complex holds the reactants in close proximity to each other, thereby eliminating the time ordinarily required for them to find each other and collide. The pump laser initiates the reaction by photolyzing the HI component of the precursor molecule. The probe laser, tuned to a resonance wavelength of the free OH, detects product formation. The laser-induced fluorescence signal of the OH shows a delay of several picoseconds due to the formation and decay of the HOCO complex. The figure below shows the energetics of the overall reaction. [Adapted from N. F. Scherer, L. R. Khundkar, R. B. Bernstein, and A. H. Zewail, *J. Chem. Phys.*, **87**, 1451 (1987).]

Transition state

H + CO_2

HOCO

HO + CO

many orders of magnitude slower than the femtoseconds to picoseconds needed to traverse the transition state. Even for crossed molecular beam experiments under the most favorable conditions, there is an uncertainty of many nanoseconds as to when the reactants will collide with one another. So there appears to be no hope of determining the

time of formation and decay of transient collision complexes with lifetimes of only femtoseconds to picoseconds.

However, a special trick enables the starting time of the reaction,
that is the zero of time, to be established within an uncertainty governed only by the duration of the laser pulse, for a whole class of bimolecular reactions. This now has opened up the possibility of studying real-time dynamics of bimolecular reactions.

The first of these experiments has been conducted at Caltech in the
picosecond, rather than femtosecond, time domain, but the same principle applies for femtochemistry. The method uses a beam of a van der
Waals "precursor molecule" containing the potential reagents in close
proximity, as prescribed by Wittig, Reisler, and coworkers at USC, and
by Benoit Soep and coworkers at the University of Paris in France.
These groups obtained the first studies of product state distributions
from reactions within the complexes.

In the picosecond experiment done in a collaboration between Bernstein and Zewail's group, the birth of OH from the reaction of atomic
hydrogen with carbon dioxide was observed:

$$H + OCO \rightarrow [HOCO]^{\neq} \rightarrow OH + CO.$$

The precursor molecule, $IH \cdots OCO$, was formed in a free-jet expansion
of a mixture of HI and CO_2 in an excess of helium carrier gas. As
pointed out by Wittig, such van der Waals molecules have favorable
geometry that limits the range of impact parameters and angles of
attack of the H on the OCO. The geometries of many such van der
Waals complexes have been determined accurately by William Klem-
perer and coworkers at Harvard University, using molecular beam
electric resonance techniques.

An ultraviolet-laser picosecond pulse initiates the experiment by
photodissociating the HI, ejecting a translationally hot H atom in the
general direction of the nearest O atom in the CO_2. A probe laser pulse
tuned to a wavelength suitable for detection of OH is delayed relative
to the photolysis pulse by specified time increments from zero to several tens of picoseconds. The OH appears after a few picoseconds. This
delay time is a direct measure of the lifetime of the HOCO collision
complex, formed from the H + CO_2 reaction, at the given relative initial translational energy. The lifetime changes with translational energy. A new series of experiments using femtosecond pulses will follow

the formation and decay of the collision complex in the femtosecond region and will be compared with predictions, such as the classical trajectory calculations by the group of George C. Schatz at Northwestern University, based on the *ab initio* calculated potential energy surfaces made by Larry Harding at Argonne National Laboratory.

It is probable that this technique will be applicable to a wide variety of bimolecular reactions, including not only those with transition states that last for picoseconds, but also direct-mode reactions with transition state lifetimes in the femtosecond range. As with unimolecular reactions, off-resonance detection should allow detailed measurements of the buildup and decay of the transition state with the zero of time for the bimolecular reaction known to femtosecond resolution.

Femtochemistry and Oriented Molecules

Connections can be made between real-time experiments on the dissociation of the photoexcited states and experiments dealing with the spatial distributions of the photoproducts. In the traditional spatial experiments, the angular distribution of the photofragments is measured by photolyzing randomly oriented ground-state molecules with linearly polarized light. The intensity of the ejected photofragments varies with the angle between the electric vector of the polarized radiation and the outgoing velocity vector of a photofragment. For a number of molecules, including CH_3I, these angular distribution experiments imply dissociation lifetimes of the excited states of less than a picosecond. For CH_3I, this promptness has been checked by direct real-time measurements, which gave a value of 0.4 picosecond or less.

A new spatial technique, introduced by the Bernstein group at UCLA, makes use of a beam of oriented symmetric top molecules prepared by an electrostatic hexapole. Orientation of the molecules means not only that one can distinguish "heads" from "tails," but that the experimenter can cause them to be arranged "heads-up" or "heads-down" at will. Polarized laser-induced photofragmentation of the oriented molecules within an electric field is carried out, and the ratio of the intensity of upward- to downward-ejected photofragments is measured. Strong up/down asymmetries are observed for several iodo-alkanes, confirming definitively the promptness of the photodissociation.

Pump-probe experiments show photofragmentation of CH₃I takes less than a picosecond

Potential energy

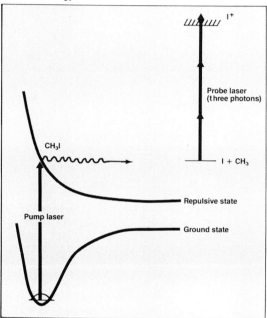

I, CH₃, separation distance

Concept of a picosecond laser experiment to determine the bond breaking time for methyl iodide is shown schematically above. The pump laser excites the CH₃I molecule from the ground state to a higher energy repulsive state. Iodine product is detected by a delayed probe pulse using three-photon ionization. Varying the delay time between the pump and probe lasers produces the results shown at right. Analysis gives an upper limit of 0.4 ps for the time required for separation of the photofragments. [Adapted from J. L. Knee, L. R. Khundkar, and A. H.

Iodine ion signal

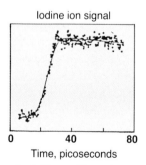

Time, picoseconds

Zewail, *J. Chem. Phys.*, **83**, 1996 (1985); L. R. Khundkar and A. H. Zewail, *Chem. Phys. Lett.*, **142**, 426 (1987).]

Photofragmentation of oriented molecules confirms prompt dissociation

Orientation field plate

I

−

Beam of oriented CH$_3$I molecules

Orientation field plate

CH$_3$

+

Vertically polarized laser beam

"Snapshot" of polarized laser fragmentation experiment on oriented CH$_3$I molecules shows, above, the molecules passing through the laser beam with the I end up. Since the I photoproduct is found to be ejected primarily in the upward direction and the CH$_3$ downward, the excited (repulsive) state must be so short-lived that the molecules do not lose their orientation. Typical experimental results are shown at right. The first arriving peak represents iodine atoms that were ejected in the upward direction from the oriented beam; the second peak comes from the downward-directed I, which takes a longer time to arrive at the detector. The degree of orientation can be estimated by comparing the intensity

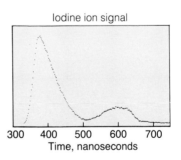

Iodine ion signal

300   400   500   600   700
Time, nanoseconds

of the two peaks. This experiment confirms that the lifetime of the transition state is less than a picosecond. [The experiment is that of S. R. Gandhi, T. J. Curtiss, and R. B. Bernstein, *Phys. Rev. Lett.*, **59**, 2951 (1987).]

Photofragmentation of oriented molecule beams clearly yields independent spatial information, complementary to the traditional angular distribution measurements, on the promptness of the decay of the photoexcited state. Moreover, the ability to select the rotational quantum states of the molecules opens up the possibility of studying the influence of rotation on the lifetime of the excited state. New theoretical advances will be required, since the role of the orientational quantum number in photoexcitation and subsequent photofragmentation of the excited state has not yet been established. Interesting quantum interference and coherence effects may be observable, as discussed by Stuart A. Rice, Shaul Mukamel, Robin Hochstrasser, and Zewail in the U.S., Moshe Shapiro in Israel, and Paul W. Brumer in Canada.

For bimolecular reactions, crossed-beam reactive scattering experiments with oriented reagent molecules confirm the shortness of the life of the collision complex in direct mode reactions. In the future, femtochemistry may be extended to study the stereochemical dynamics of these reactions in real time.

FEMTOCHEMISTRY AND RESONANCES

The femtochemistry discussed thus far has dealt with elementary chemical reactions, such as $ICN^* \rightarrow [I \cdots CN]^{\neq *} \rightarrow I + CN$, for which bond breaking takes place on dissociative potential surfaces. The molecules are in "transition" for only a few hundred femtoseconds or so, as evidenced by the rise and decay observed in the ICN experiments and confirmed by theory. If, however, in the process of falling apart, the system encounters a well in the potential surface, or if there is more than one degree of freedom involved, the system can be "trapped" and thus exhibit behavior indicative of quasi-bound states, or resonances. Reactive scattering resonances were predicted based on theory in the early 1970s by Raphael D. Levine of Hebrew University of Jerusalem, Aron Kuppermann at Caltech, and Donald G. Truhlar, now at the University of Minnesota.

Manifestation of these resonances in the real-time probing of fragment separation would be a slowdown in the appearance of free fragments and possibly the appearance of oscillations reflecting the vibrational resonance frequency of the wave packet of the dissociating fragments. The simplest case would be a diatomic salt molecule, MX (from the alkali age), for which the initially excited state, MX*, is covalent. This state interacts with the ground electronic state, which is ionic (M^+X^-). Then the dynamics of bond breaking may involve the phenomenon termed "potential curve crossing" as a result of this covalent–ionic interaction.

In a classic series of papers dating from the 1960s, R. Stephen Berry and coworkers at the University of Chicago provided the foundation for the description of the alkali halide potentials and the avoided crossing which defines the adiabatic (trapped MX*) and diabatic (MX* → M + X) behavior of the excited salt molecule. En route to products, the $[M \cdots X]^{\neq*}$ transition-state molecules "decide" between the covalent and ionic potentials. Either the wave packet of the MX* will be trapped on the adiabatic potential curve without crossing, or it will dissociate by following the diabatic curve. The two limits have entirely different temporal behavior and, if there is trapping, the frequency and amplitude of the oscillations will provide details of the nature of the surfaces and the strength of the coupling. Roger Grice at the University of Manchester in the U.K. and Herschbach have provided a theory for this coupling in the Landau–Zener regimes for alkali halides, and Mark S. Child at Oxford University has treated the role of the coupling in determining the lifetime of quasi-bound states. These systems have been studied experimentally by absorption, emission, and molecular beam photofragmentation techniques by a number of researchers.

Such trapping resonances have been observed in femtochemistry experiments at Caltech on the reactions

$$NaI^* \rightarrow [Na \cdots I]^{\neq*} \rightarrow Na + I$$

and

$$NaBr^* \rightarrow [Na \cdots Br]^{\neq*} \rightarrow Na + Br.$$

For NaI, the oscillations are very strong and the average period is 1.25 picoseconds, which corresponds to a frequency of 27 cm^{-1}. This leads to the conclusion that the packet for NaI is effectively trapped in the adiabatic well en route to products, and that the crossing is inefficient. The experiments show that the oscillation damping time for NaI is about 10 picoseconds. Thus, one crossing on the outward phase per oscillation (about 1 picosecond) has a probability of about 0.1 to escape

Femtochemistry reveals fragment trapping in alkali halide photodissociation

Potential energy

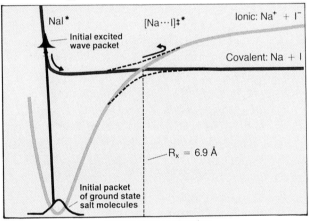

Internuclear separation

Wave packet description of the photofragmentation of sodium iodide with two electronic degrees of freedom is shown schematically above. Light excites a packet of ground state NaI molecules into an initial excited state wave packet, shown as the upper bell-shaped curve in the diagram. From here the molecules may follow the purple diabatic potential curve and dissociate covalently, or they may attempt to cross to the ionic curve at an internuclear distance of 6.9 Å. Molecules that attempt this crossing become trapped and oscillate back and forth in the upper potential energy well. Such trapping can be seen experimentally on the femtosecond timescale, at right. The red curve shows the Na—I bond resonating in trapped molecules. The blue curve is that of the sodium

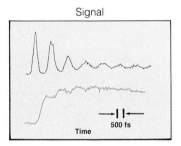

Signal

when separating from the iodine along the covalent curve. The curves were observed with different degrees of detuning. [Adapted from M. Rosker, T. S. Rose, and A. H. Zewail, *Chem. Phys. Lett.*, **146**, 175 (1988) and T. S. Rose, M. Rosker, and A. H. Zewail, *J. Chem. Phys.*, **88**, 6672 (1988).]

from the well. For NaBr, the frequency of oscillation is similar in magnitude, but severe damping is observed. Thus, the crossing for NaBr is much easier than for NaI, consistent with theoretical expectation.

An interesting question is how the "dephasing" and spread of the wave packet influences the dissociation rate, as determined by the decay of the oscillations. The experiments show manifestations of this decay and spreading, and theoretical modeling of the wave packet dynamics in these prototype systems will be important. Very recently Horia I. Metiu of the University of California, Santa Barbara, Marcus, Zewail, and coworkers have shown that this experimental resonance behavior can be reproduced by a quantum calculation.

These real-time observations of trapping resonance allow the motion of the wave packet to be viewed from the initial excitation all the way to "infinite separation," or fragmentation to atoms, making contact with absorption spectroscopy, atomic beam scattering experiments, and photofragment translational spectroscopy. Such experiments promise to provide rich dynamical information bearing on the shape of the potential energy surfaces, curve crossings, and interactions among different degrees of freedom.

PERSPECTIVES

It has now been 40 years since the introduction to photochemistry of flash photolysis with a time scale of microseconds. Since then the pulse duration has been continually improved. In laser femtochemistry, the "shutter speed" has been increased by nearly nine orders of magnitude: The present limit of 6-fs time resolution corresponds to a spatial window of about 0.06 Å through which to view the potential energy surface.

But more important, with the last three orders of magnitude, it has become possible to observe chemistry as it happens. Femtochemistry is bringing reality to the ephemeral, but all-important, transition states in chemical reactions. This happy marriage of ultrafast lasers and chemistry promises an exciting future for this field of real-time molecular reaction dynamics.

SUGGESTED READINGS

Ashfold, M. N. R., and Baggott, J. E., eds. (1987). *Molecular Photodissociation Dynamics,* Advances in Gas-Phase Photochemistry and Kinetics. Royal Society of Chemistry, U.K.

Benson, S. W. (1982). *The Foundations of Chemical Kinetics.* Krieger Publishing Co., Malabar, Florida.

Bernstein, R. B., Herschbach, D. R., Levine, R. D., eds. (1987). "Dynamical Stereochemistry," *J. Phys. Chem.* **91.** (Oct. 8, 1987).

Brooks, P. R. (1988). "Spectroscopy of Transition Region Species," *Chem. Rev.* **88,** 407.

Eisenthal, K. B. (1975). "Studies of Chemical and Physical Processes with Picosecond Lasers," *Acc. Chem. Res.* **8,** 118.

Eyring, H., Lin, S. H., and Lin, S. M. (1980). *Basic Chemical Kinetics.* John Wiley & Sons, New York.

Felker, P. M., and Zewail, A. H. (1988). "Picosecond Dynamics of IVR," *Adv. Chem. Phys.* **70,** 265.

Fleming, G. R., and Siegman, A. E., eds. (1986). *Ultrafast Phenomena V.* Springer-Verlag, New York.

Herschbach, D. R. (1987). "Molecular Dynamics of Elementary Chemical Reactions," *Les Prix Nobel in 1986.* Elsevier Publishing Co., Amsterdam.

Hochstrasser, R. M., and Weisman, R. B. (1980). "Picosecond Relaxation of Electronically Excited Molecular States in Condensed Media," in *Radiationless Transitions* (S. H. Lin, ed.), p. 317. Academic Press, New York.

Jortner, J., Levine, R. D., and Rice, S. A. (1981). "Photoselective Chemistry," *Advances in Chemical Physics,* Vol. XLVII, Pt. 1 & 2. John Wiley & Sons, New York.

Kaufmann, K. J., and Rentzepis, P. M. (1975). "Picosecond Spectroscopy in Chemistry and Biology," *Acc. Chem. Res.* **8,** 408.

Knee, J. L., and Zewail, A. H. (1988). "Ultrafast Laser Spectroscopy of Chemical Reactions," *Spectroscopy* **3,** 44.

Laubereau, A., and Kaiser, W. (1978). "Vibrational Dynamics of Liquids and Solids Investigated by Picosecond Light Pulses," *Rev. Mod. Phys.* **50,** 607.

Lee, Y. T. (1987). "Molecular Beam Studies of Elementary Chemical Processes," *Science* **236,** 793.

Letokhov, V. S. (1983). *Nonlinear Laser Chemistry,* Springer Series in Chemical Physics 22. Springer-Verlag, New York.

Levine, R. D., and Bernstein, R. B. (1987). *Molecular Reaction Dynamics and Chemical Reactivity.* Oxford University Press, New York.

Marcus, R. A. (1977). "Energy Distributions in Unimolecular Reactions," *Ber. Bunsenges. Phys. Chem.* **81,** 190.

Pilling, M. J., Smith, I. W. M., eds. (1987). *Modern Gas Kinetics.* Blackwell Scientific Publications, Oxford.

Polanyi, J. C. (1987). "Some Concepts in Reaction Dynamics," *Science* **236,** 680.

Shank, C. V. (1986). "Investigation of Ultrafast Phenomena in the Femtosecond Time Domain," *Science* **233,** 1276.

Siegman, A. E. (1986). *Lasers.* University Science Books, Mill Valley, California.

Smith, I. W. M. (1987). "Direct Probing of Reactions," *Nature* **328,** 760.

Tannor, D. J., and Rice, S. A. (1988). "Coherent Pulse Sequence Control of Product Formation in Chemical Reactions," *Adv. Chem. Phys.* **70,** 441.

Truhlar, D. G. (1984). *Resonances,* ACS Symposium Series 263. American Chemical Society, Washington, D.C.

Zare, R. N., and Dagdigian, P. J. (1974). *Science* **185,** 739.

Zewail, A. H., Letokhov, V. S., Zare, R. N., Bernstein, R. B., Lee, Y. T., and Shen, Y. R. (1980). *Physics Today* **33**(11) (November 1980).

"Dynamics of Molecular Photofragmentation" (1986). *Faraday Disc. Chem. Soc.* **82.**

AN UPDATE

Since the writing of this article on femtochemistry, a number of developments has taken place. Scientifically, these developments and the progress in this field have continued to meet expectations. On the personal side, however, we all suffered the tragic loss of Dick Bernstein,the co-author and a dear friend, who died on July 8, 1990, in Helsinki. Dick had the wisdom and vision to see the trajectory of femtochemistry, and to him I dedicate this updated chapter—I know how happy he would have been with the recent progress. (The chapter represents the contents of the talk that Dick was to give at the conference.)

Over the past five years, there have been more than 400 papers published in relation to experimental and theoretical femtochemistry [*Science Watch* **2** (6), 3–5 (1991)]. The picosecond work on vibrational and rotational energy redistribution and the state-to-state rates are not included. Here, we shall give references to the systems which have been experimentally studied so far. [For review see L. Khundkar and A. H. Zewail, *Ann. Rev. Phys. Chem.* **41**, 15–60 (1990).]

In the article of this chapter, reference was made to the possibility of observing the collision complex of a bimolecular reaction. This has been achieved for the reaction of a Br atom with an I_2 molecule. Rydberg states femtosecond dynamics have also been studied using multi-photon–mass spectrometric detection methods. The wave packet dynamics of metal clusters, Na_2 and Na_3, have been observed. The dynamics of transition states at saddle-points have been studied in real-time for the $[HgI_2]^{\ddagger*}$ system. Heavier molecular systems, Bi_2, have been studied using subpicosecond transient absorption techniques and femtosecond transition-state spectroscopy. Finally, new observations were made in the NaI femtosecond experiments.

The limit of the uncertainty principle for femtosecond pulses was addressed in a series of experiments aimed at probing the wave packet motion in bound potential systems, I_2 and ICl molecules. It was shown that one can recover all information on the vibrational/rotational motion if the coherence of the laser is exploited. In other words, broad-energy pulses can detect sharp resonances. With this later advance, femtochemistry has now covered the following cases: motion on bound potentials, direct bond breakage on repulsive potentials, covalent-to-ionic bond dynamics (caused by avoided curve crossing), reactions involving more than one degree of vibrational freedom (potential energy surfaces with a saddle point), high-energy predissociating Rydberg states (involving bound and repulsive potentials), and the collision complex dynamics in bimolecular reactions involving potential wells. There are some new studies and directions which are discussed in the articles given below.

REVIEW AND INTRODUCTORY ARTICLES

Smith, I. W. M. (1987). "Direct probing of reactions," *Nature* **328,** 760–761.

Zewail, A. H. (1988). "Laser femtochemistry," *Science* **242,** 1645–1653.

Knee, J. L., and Zewail, A. H. (1988). "Ultrafast laser spectroscopy of chemical reactions," *Spectroscopy* **3,** 44–53.

Zewail, A. H. (1989). "Femtochemistry: The role of alignment and orientation," *J. Chem. Soc. Faraday Trans. 2* **85,** 1221–1242.

Khundkar, L. R., and Zewail, A. H. (1990). "Ultrafast molecular reaction dynamics in real-time: Progress over a decade," *Ann. Rev. Phys. Chem.* **41,** 15–60.

Gruebele, M., and Zewail, A. H. "Ultrafast reaction dynamics." (1990): *Phys. Today* **43**(5), 24–33; *Ber. Bunsenges. Phys. Chem.* **94,** 1210 (in German); *Parity* **5**(12), 8 (in Japanese). (1991): *Sov. Phys. Uspekhi* **161**(3), 69 (in Russian).

Zewail, A. H. (1990). "The birth of molecules," *Scientific American* **263**(6), 76–82. Translated in *Le Scienze* (Italian), *Saiensu* (Japanese), *Pour la Science* (French), *Investigacion y CIENCIA* (Spanish), *Spectrum der Wissenschaft* (German), *V Mire Nauki* (Russian), *Ke Xue* (Chinese), *Majallat ALOLOOM* (Arabic), *Tudomany* (Hungarian), *Vigyan* (Indian), and *AlKhairia* (English), *Proceedings of the Royal Institution* (United Kingdom).

Smith, I. W. M. (1990). "Exposing molecular motions," *Nature* **343,** 691–692.

Zewail, A. H. (1991). "Femtosecond transition-state dynamics," *Faraday Discuss. Chem. Soc.* **91,** 207–237.

Dantus, M., and Roberts, G. (1991). "Femtosecond transition-state spectroscopy and chemical reaction dynamics," *Comments Atom. Molec. Phys.* **26,** 131–152.

Thomas, J. M. (1991). "Femtosecond diffraction," *Nature* **351,** 694–695.

Atkins, P. W. (1990). "Physical Chemistry," 4th Ed., p. 856. Freeman Publishing Co.

Brown, T. L., LeMay, H. E., Jr., and Bursten, B. E. (1991). "Chemistry: The Central Science," 5th Ed., p. 495. Prentice-Hall, New Jersey.

Asimov, Isaac (1991). "Breaking the bond," *in* "Frontiers." Truman Talley Books/Plume, New York.

Yoshihara, K. (1991). "Ultrafast Molecular Dynamics" Oyo Butsuri (*App. Phys.*) **60,** 891–898.

DIRECT BOND BREAKAGE (REPULSIVE SURFACES)
ICN—EXPERIMENTAL AND THEORY

Scherer, N. F., Knee, J. L., Smith, D. D., and Zewail, A. H. (1985). "Femtosecond photofragment spectroscopy: The reaction ICN → CN + I," *J. Phys. Chem.* **89,** 5140–5143.

Dantus, M., Rosker, M. J., and Zewail, A. H. (1987). "Real-time femtosecond probing of 'transition states' in chemical reactions," *J. Chem. Phys.* **87,** 2395–2397.

Rosker, M. J., Dantus, M., and Zewail, A. H. (1988). "Femtosecond clocking of the chemical bond," *Science* **241,** 1200–1202.

Rosker, M. J., Dantus, M., and Zewail, A. H. (1988). "Femtosecond real-time probing of reactions I: The technique," *J. Chem. Phys.* **89,** 6113–6127.

Dantus, M., Rosker, M. J., and Zewail, A. H. (1987). "Femtosecond real-time probing of reactions II: The dissociation reaction of ICN," *J. Chem. Phys.* **89,** 6128–6140.

Dantus, M., Bowman, R. M., Baskin, J. S., and Zewail, A. H. (1989). "Femtosecond real-time alignment in chemical reactions," *Chem. Phys. Lett.* **159**, 406–412.

Bersohn, R., and Zewail, A. H. (1988). "Time dependent absorption of fragments during dissociation," *Ber. Bunsenges. Phys. Chem.* **92**, 373–378.

Williams, S. O., and Imre, D. G. (1988). "Time evolution of single- and two-phonon processes for a pulse-mode laser," *J. Phys. Chem.* **92**, 6636–6647.

Williams, S. O., and Imre, D. G. (1988). "Determination of real time dynamics in molecules by femtosecond laser excitation," *J. Phys. Chem.* **92**, 6648–6654.

Bernstein, R. B., and Zewail, A. H. (1989). "Femtosecond real-time probing of reactions III: Inversion to the potential from femtosecond transition-state spectroscopy experiments," *J. Chem. Phys.* **90**, 829–842.

Zewail, A. H. (1989). "Femtochemistry: The role of alignment and orientation," *J. Chem. Soc., Faraday Trans. 2* **85**, 1221–1242.

Dantus, M., Bowman, R. M., Baskin, J. S., and Zewail, A. H. (1989). "Femtosecond real-time alignment in chemical reactions," *Chem. Phys. Lett.* **159**, 406–412.

Heather, R., and Metiu, H. (1989). "Rotational coherence effects in the femtosecond photodissociation of ICN," *Chem. Phys. Lett.* **160**, 531–537.

Lee, S.-Y., Pollard, W. T., and Mathies, R. A. (1989). "Quantum theory for transition state absorption," *Chem. Phys. Lett.* **160**, 531–537.

Benjamin, I., and Wilson, K. R. (1989). "Proposed experimental probes of chemical reaction molecular dynamics in solution: ICN photodissociation," *J. Chem. Phys.* **90**, 4176–4197.

Henriksen, N. E., and Heller, E. J. (1989). "Quantum dynamics for vibrational and rotational degrees of freedom using Gaussian wavepackets: Application to the three-dimensional photodissociation dynamics of ICN," *J. Chem. Phys.* **91**, 4700–4713.

Yan, Y. J., Fried, L. E., and Mukamel, S. (1989). "Ultrafast pump-probe spectroscopy: Femtosecond dynamics in Liouville space," *J. Phys. Chem.* **93**, 8149–8162.

Fried, L. E., and Mukamel, S. (1990). "A classical theory of pump-probe photodissociation for arbitrary pulse durations," *J. Chem. Phys.* **93**, 3063–3071.

Mukamel, S. (1990). "Femtosecond optical spectroscopy: A direct look at elementary chemical events," *Ann. Rev. Phys. Chem.* **41**, 647–681.

Beswick, J. A., and Jortner, J. (1990). "Time scales for molecular photodissociation," *Chem. Phys. Lett.* **168**, 246–248.

Krause, J. L., Shapiro, M., and Bersohn, R. (1991). "Derivation of femtosecond pump-probe dissociation transients from frequency resolved data," *J. Chem. Phys.* **94**, 5499–5507.

Roberts, G., and Zewail, A. H. (1991). "Femtosecond real-time probing of reactions VII: A quantum and classical mechanical study of the ICN dissociation experiment," *J. Phys. Chem.* **95**, 7973–7993.

COVALENT-TO-IONIC MOTION (AVOIDED CURVE CROSSING) NaI—EXPERIMENTAL AND THEORY

Rose, T. S., Rosker, M. J., and Zewail, A. H. (1988). "Femtosecond real-time observation of wavepacket oscillations (resonance) in dissociation reactions," *J. Chem. Phys.* **88**, 6672–6673.

Rosker, M. J., Rose, T. S., and Zewail, A. H. (1988). "Femtosecond real-time dynamics of photofragment-trapping resonances on dissociative potential energy surfaces," *Chem. Phys. Lett.* **146**, 175–179.

Rose, T. S., Rosker, M. J., and Zewail, A. H. (1989). "Femtosecond real-time probing of reactions IV: The reactions of alkali halides," *J. Chem. Phys.* **91**, 7415–7436.

Cong, P., Mokhtari, A., and Zewail, A. H. (1990). "Femtosecond probing of persistent wavepacket motion in dissociative reactions: Up to 40 ps," *Chem. Phys. Lett.* **172**, 109–113.

Mokhtari, A., Cong, P., Herek, J. L., and Zewail, A. H. (1990). "Direct femtosecond mapping of trajectories in a chemical reaction," *Nature* **348**, 225–227.

Engel, V., Metiu, H., Almeida, R., Marcus, R. A., and Zewail, A. H. (1988). "Molecular state evolution after excitation with an ultra-short laser pulse: A quantum analysis of NaI and NaBr dissociation," *Chem. Phys. Lett.* **152**, 1–7.

Marcus, R. A. (1988). "Semiclassical wavepackets in the angle representation and their role in molecular dynamics," *Chem. Phys. Lett.* **152**, 8–13.

Rose, T. S., Rosker, M. J., and Zewail, A. H. (1989). "Femtosecond real-time probing of reactions IV: The reactions of alkali halides," *J. Chem. Phys.* **91**, 7415–7436.

Engel, V., and Metiu, H. (1989). "The study of NaI predissociation with pump-probe femtosecond laser pulses: The use of an ionizing probe pulse to obtain more detailed information," *Chem. Phys. Lett.* **155**, 77–82.

Lin, S. H., and Fain, B. (1989). "Application of the theory of two-dimensional spectroscopy to the real-time femtosecond transition state spectroscopy," *Chem. Phys. Lett.* **155**, 216–220.

Choi, S. E., and Light, J. C. (1989). "Use of the discrete variable representation in the quantum dynamics by a wavepacket propagation: Predissociation of $NaI(1\Sigma_0^+) \rightarrow NaI(O+) \rightarrow Na(^2S) + I(^2P)$," *J. Chem. Phys.* **90**, 2593–2604.

Engel, V., and Metiu, H. (1989). "A quantum mechanical study of predissociation dynamics of NaI excited by a femtosecond laser pulse," *J. Chem. Phys.* **90**, 6116–6128.

Lee, S.-Y., Pollard, W. T., and Mathies, R. A. (1989). "Classical theory for real-time femtosecond probing of the NaI* photodissociation," *J. Chem. Phys.* **90**, 6149–6150.

Engel, V., and Metiu, H. (1989). "Two-photon excitation of NaI with femtosecond laser pulses," *J. Chem. Phys.* **91**, 1596–1602.

Fain, B., Lin, S. H., and Hamer, N. (1989). "Two-dimensional spectroscopy: Theory of nonstationary, time-dependent absorption and its application to femtosecond processes," *J. Chem. Phys.* **91**, 4485–4494.

Wang, J., Blake, A. J., McCoy, D. G., and Torop, L. (1990). "Analytical potential curves for the $X^1\Sigma+$ and $O+$ states of NaI," *Chem. Phys. Lett.* **175**, 225–230.

Chapman, S., and Child, M. S. (1991). "A semiclassical study of long time recurrences in the femtosecond predissociation dynamics of NaI," *J. Phys. Chem.* **95**, 578–584.

Yamashita, K., and Morokuma, K. (1991). "Ab initio molecular orbital and dynamics study of transition-state spectroscopy," *Faraday Discuss. Chem. Soc.* **91**.

Kono, H., and Fujimura, Y. (1991). "Interference between nonadiabatically ramified wave packets in dissociation dynamics of NaI excited by a femtosecond laser pulse," *Chem. Phys. Lett.* **184**, 497.

Meier, Ch., Engel, V., and Briggs, J. S. (1991). "Long time wave packet behavior in a curve-crossing system: The predissociation of NaI." *J. Chem. Phys.* **95**, 7337–7343.

BOUND SYSTEMS (BOUND POTENTIALS)
I_2 AND ICl—EXPERIMENTAL AND THEORY

Bowman, R. M., Dantus, M., and Zewail, A. H. (1989). "Femtosecond transition-state spectroscopy of iodine: From strongly bound to repulsive surface dynamics," *Chem. Phys. Lett.* **161**, 297–302.

Dantus, M., Bowman, R. M., and Zewail, A. H. (1990). "Femtosecond laser observations of molecular vibration and rotation," *Nature* **343**, 737–739.

Bernstein, R. B., and Zewail, A. H. (1990). "From femtosecond temporal spectroscopy to the potential by a direct classical inversion method," *Chem. Phys. Lett.* **170**, 321–328.

Gruebele, M., Roberts, G., Dantus, M., Bowman, R. M., and Zewail, A. H. (1990). "Femtosecond temporal spectroscopy and direct inversion to the potential: Application to iodine," *Chem. Phys. Lett.* **166**, 459–469.

Gerdy, J. J., Dantus, M., Bowman, R. M., and Zewail, A. H. (1990). "Femtosecond selective control of wavepacket population," *Chem. Phys. Lett.* **171**, 1–4.

Bowman, R. M., Dantus, M., and Zewail, A. H. (1990). "Femtosecond multiphonon dynamics of higher-energy potentials," *Chem. Phys. Lett.* **174**, 546–552.

Janssen, M. H. M., Bowman, R. M., and Zewail, A. H. (1991). "Femtosecond temporal spectroscopy of ICl: Inversion to the $A^3\pi_1$ state potential," *Chem. Phys. Lett.* **172**, 99–108.

Dantus, M., Janssen, M. H. M., and Zewail, A. H. (1991). "Femtosecond probing of molecular dynamics by mass-spectrometry in a molecular beam," *Chem. Phys. Lett.* **181**, 281–287.

Scherer, N. F., Ruggiero, A. J., Du, M., and Fleming, G. R. (1990). "Time resolved dynamics of isolated molecular systems studied with phase-locked femtosecond pulse pairs," *J. Chem. Phys.* **93**, 856–857.

Scherer, N. F., Carlson, R. J., Matro, A., Du, M., Ruggiero, A. J., Romero-Rochin, V., Cina, J. A., Fleming, G. R., and Rice, S. A. (1991). "Fluorescence-detected wavepacket intereferometry: Time resolved molecular spectroscopy with sequences of femtosecond phase-locked pulses," *J. Chem. Phys.* **95**, 1487–1511.

Letokhov, V. S., and Tyakht, V. V. (1990). "Coherent vibrational wavepacket dynamics in femtosecond laser excitation of diatomic molecules," *Israel J. Chem.* **30**, 189–195.

Metiu, H., and Engel, V. (1990), "A theoretical study of I_2 vibrational motion after excitation with an ultrashort pulse," *J. Chem. Phys.* **93**, 5693–5699.

Hartke, B. (1990). "Quantum simulation of femtosecond population control in iodine," *Chem. Phys. Lett.* **175**, 322–326.

SADDLE-POINT TRANSITION STATES (POTENTIAL ENERGY SURFACES)
HgI$_2$—EXPERIMENTAL AND THEORY

Bowman, R. M., Dantus, M., and Zewail, A. H. (1989). "Femtochemistry of the reactions: IHgI*→[IHg···I]‡*→HgI + I," *Chem. Phys. Lett.* **156,** 131–137.

Dantus, M., Bowman, R. M., Gruebele, M., and Zewail, A. H. (1989). "Femtosecond real-time probing of reactions V: The reaction of IHgI," *J. Chem. Phys.* **91,** 7437–7450.

Dantus, M., Bowman, R. M., Baskin, J. S., and Zewail, A. H. (1989). "Femtosecond real-time alignment in chemical reactions," Chem. Phys. Lett. **159,** 406–412.

Gruebele, M., Roberts, G., and Zewail, A. H. (1990). "Femtochemistry of the reaction of IHgI: Theory versus experiment," *Philos. Trans. Roy. Soc. A (London)* **332,** 223–243.

DIRECT AND RYDBERG-STATE PREDISSOCIATION
(BOUND + REPULSIVE)
CH$_3$I—EXPERIMENTAL

Knee, J.L., Khundkar, L. R., and Zewail, A. H. (1985). "Picosecond monitoring of a chemical reaction in molecular beams: Photofragmentation of R–I→R$\ddagger$ + I," *J. Chem. Phys.* **83,** 1996–1998.

Khundkar, L. R., and Zewail, A. H. (1987). "Picosecond MPI mass spectroscopy of CH$_3$I in the process of dissociation," *Chem. Phys. Lett.* **142,** 426–432.

Dantus, J., Janssen, M. H. M., and Zewail, A. H. (1991). "Femtosecond probing of molecular dynamics by mass-spectrometry in a molecular beam," *Chem. Phys. Lett.* **181,** 281–287.

[There are related theoretical studies by J. Kinsey, G. Schatz, L. Butler, L. Zeigler, V. Vaida, E. Grant, R. Dixon and others, but studies of the theoretical femtochemistry dynamics of the A-continuum and high-energy Rydberg states are still in progress.]

CLASSICAL SYSTEMS (SLOWER MOTION ON REPULSIVE SURFACES)
BI₂—EXPERIMENTAL AND THEORY

Misewich, J. A., Glownia, J. H., Walkup, R. E., and Sorokin, P. P. (1990). "Femtosecond transition-state absorption spectroscopy of Bi atoms produced by Bi_2 photodissociation," *in* "Ultrafast Phenomena VII" (C. B. Harris, E. P. Ippen, G. A. Mourou, and A. H. Zewail, eds.), Springer Series in Chemical Physics, Vol. 53, pp. 426–428. Springer-Verlag, Berlin.

Glownia, J. H., Misewich, J. A., and Sorokin, P. P. (1990). "Femtosecond transition-state absorption spectroscopy of Bi atoms produced by photodissociation of gaseous Bi_2 molecules," *J. Chem. Phys.* **92,** 335–3339.

Walkup, R. E., Misewich, J. A., Glownia, J. H., and Sorokin, P. P. (1990). "Time-resolved absorption spectra of dissociating molecules," *Phys. Rev. Lett.* **65,** 2366–2369.

Walkup, R. E., Misewich, J. A., Glownia, J. H., and Sorokin, P. P. (1991). "Classical model of time-resolved absorption spectra of dissociating molecules," *J. Chem. Phys.* **94,** 3389–3406.

Bowman, R. M., Gerdy, J. J., Roberts, G., and Zewail, A. H. (1991). "Femtosecond real-time probing of reactions VI: A joint experimental and theoretical study of Bi_2 dissociation," *J. Phys. Chem.* **95,** 4635–4647.

METALLIC CLUSTERS (IONIZATION AND FRAGMENTATION)
Na₂ AND Na₃—EXPERIMENTAL AND THEORY

Baumert, T., Bühler, B., Thalweiser, R., and Gerber, G. (1990). "Femtosecond spectroscopy of molecular autoionization and fragmentation," *Phys. Rev. Lett.* **64,** 733–736.

Baumert, T., Bühler, B., Grosser, M., Thalweiser, R., Weiss, V., Wiedenmann, E., and Gerber, G. (1991). "Femtosecond time-resolved wave packet motion in molecular multiphoton ionization and fragmentation," *J. Phys. Chem.* **95,** 8103–8110.

Kowalczyk, P., Radzewicz, C., Mostowski, J., and Walmsley, I. A. (1990). "Time-resolved luminescence from coherently excited molecules as a probe of molecular wavepacket dynamics," *Phys. Rev. A* **42,** 5622–5626.

Engel, V. (1991). "Femtosecond pump/probe experiments and ionization: The time dependence of the total ion signal," *Chem. Phys. Lett.* **178,** 130–134.

BIMOLECULAR REACTIONS
$Br + I_2$ AND $IH + CO_2$—EXPERIMENTAL AND THEORY

Scherer, N. F., Khundkar, L. R., Bernstein, R. B., and Zewail, A. H. (1987)."Real-time picosecond clocking of the collision complex in a bimolecular reaction: The birth of OH from $H + CO_2$," *J. Chem. Phys.* **87,** 1451–1453.

Scherer, N. F., Sipes, C., Bernstein, R. B., and Zewail, A. H. (1990). "Real-time clocking of bimolecular reactions: Application to $H + CO_2$," *J. Chem. Phys.* **92,** 5239–5259.

Gruebele, M., Sims, I. R., Potter, E. D., and Zewail, A. H. (1991). "Femtosecond probing of bimolecular reactions: The collision complex," *J. Chem. Phys.* **95,** 7763–7766.

[There are many theoretical papers on this subject reviewed in the work of G. Schatz, L. Harding, K. Morokuma . . . (see references in the above articles) which focus on the potential energy surface and the dynamics governed by these potentials.]

Some comments related to other work discussed in the chapter are now in order. First, in the last few years, important progress has been made in the study of the $H + H_2$ family of reactions, both on the experimental side (by the groups of R. Zare, J. Valentini, Y. Lee, R. Gentry, P. Toennies, K. Welge, and others), and on the theoretical front by the groups of W. Miller, D. Truhlar, D. Kuri, R. Wyatt, J. C. Light, G. Schatz, and others (for a recent review, see 1). Second, transition-state spectroscopy, mentioned in the chapter, has now been further advanced on several systems using elegant photodetachment methods from the ion to the neutral by the group of Neumark (2), and by using the pioneering approach of neutral-to-neutral photoabsorption in vdW systems near TS geometries by the group of B. Soep (3).

A large number of systems have continued to be studied by molecular beam techniques, chemilumiscence, photofragment spectroscopy, time-integrated transition-state spectroscopy, state-to-state rates, and product-state distributions, but we limited the "update" here mostly to femtochemical and related studies.

1. Miller, W. H., and Zhang, J. Z. H. (1991). *J. Phys. Chem.* **95,** 12, and references therein.

2. Metz, R. B., Weaver, A., Bradforth, S. E., Kitsopoulos, T. N., and Neumark, D. M. (1990). *J. Phys. Chem.* **94,** 1377.

3. Jouvet, C., and Soep, B. (1984). *J. Chem. Phys.* **80,** 2229; see also *Faraday Discuss. Chem. Soc.* **91** (1991).

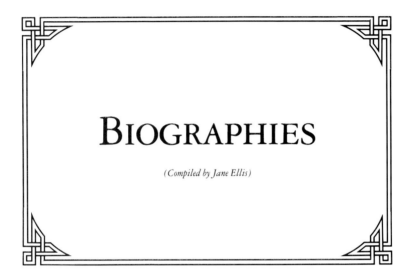

BIOGRAPHIES

(Compiled by Jane Ellis)

Linus Pauling with a model of the alpha helix, showing side chains for amino-acid residues, in his California Institute of Technology laboratory, about 1952.

Linus Pauling

———•———◇———•———

Pauling was born in Portland, Oregon on February 28, 1901. After receiving his elementary, high school, and college education in Oregon, he came to Caltech, where he obtained his Ph.D. in 1925. With the help of a Guggenheim Fellowship, he studied quantum mechanics in Europe for 18 months. He then returned as a faculty member to Caltech where he remained until 1964. During the early part of his Caltech career, his research centered on problems in structural chemistry, especially on the determination of the structures of molecules through x-ray diffraction and electron diffraction. In 1939, many of his discoveries and insights led to *The Nature of the Chemical Bond,* one of the most influential scientific books of the 20th century.

In the mid-1930s, Pauling became interested in biological molecules. His efforts included magnetic studies in oxygen-carrying hemoglobin molecules, and the development of a structural theory of denatured protein molecules. His research projects were interrupted by World War II, during which he worked on explosives and developed an oxygen detector. He continued his interest in biological molecules through his work on antibodies and on an artificial blood serum. In the late 1940s, he discovered the alpha helix as the basic structure of proteins. In 1954, he was awarded the Nobel Prize in chemistry for his outstanding contributions toward understanding chemical bonding. On October 10, 1963, Pauling, for his peace efforts, was awarded his second Nobel Prize.

Pauling's later career has concentrated on the chemistry of life and challenges in medicine. He showed, with Harvey Itano, that sickle-cell anemia is a hereditary molecular disease, and he has investigated the value of large doses of vitamins for the treatment of various diseases.

Max Perutz in about 1985.

Max F. Perutz

Max Perutz was born in Vienna and received his Ph.D. from Cambridge University in 1940. He was for many years the Director of the Medical Research Council Unit for Molecular Biology at the Cavendish Laboratory in Cambridge and later Chairman of the Medical Research Council Laboratory there. He has been the recipient of the Royal Medal of the Royal Society, the Copley Medal of the Royal Society, and the Nobel Prize for Chemistry, which was awarded to him in 1962. He became a Companion of Honour in 1975 and was awarded the Order of Merit in 1988. Dr. Perutz lives in Cambridge.

Alexander Rich, 1986.

Alexander Rich

Alexander Rich was born in Hartford, Connecticut, and educated at Harvard University. His undergraduate studies at Harvard College centered on physics, chemistry, and biochemical sciences, and he received an M.D. degree at Harvard Medical School in 1949. He then went to the chemistry department at Caltech to do postdoctoral work with Linus Pauling for the next five years. During that period he initiated x-ray diffraction studies both on small molecules and on polymeric nucleic acids. From 1954 to 1958 he was head of the Section on Physical Chemistry at the National Institutes of Health in Bethesda, Maryland. In the period 1955 to 1956 he worked as a visiting scientist at the Cavendish Laboratory in Cambridge, England. At NIH and in England, his research focused on discovering the three-dimensional structure of collagen—the fibrous molecule of skin and connective tissue—and on working out the structure and reactions of a large number of polyribonucleotides. It was during this period that the formation of triple-stranded nucleic acid complexes was discovered.

In 1958 Rich moved to the biology department at M.I.T. and continued studies both of nucleic acid structure and of molecular biology. His work included the discovery and characterization of the polyribosomal structures involved in protein synthesis (polysomes), discovery of DNA in chloroplasts, and determination of the three-dimensional structure of yeast phenylalanine transfer RNA, an important component of the protein synthesis system. X-ray diffraction work on synthetic polynucleotides led to atomic resolution determination of the structure of the nucleic acid double helix. More recent work involved discovery of the left-handed form of DNA, called Z-DNA, and characterization of those factors that influence its formation both in solution and inside cells.

Alexander Rich has served on a number of committees dealing with science policy. He has also maintained an active interest in arms control and world peace. Rich has also worked on the origin of life. He served on NASA's Lunar and Planetary Missions Board and was a member of the Viking mission team that searched for active biology on the surface of Mars. Rich is currently the William Thompson Sedgwick Professor of Biophysics in the biology department at M.I.T.

Francis Crick with two daughters, Gabrielle (left) and Josephine, circa 1956.

Francis Crick

———•———◇———•———

Dr. Francis Crick is J. W. Kieckhefer Distinguished Research Professor at The Salk Institute for Biological Studies, La Jolla, California. He is also an adjunct professor of psychology at the University of California at San Diego. He was a founder and member of the Medical Research Council's Laboratory of Molecular Biology at Cambridge from 1961 to 1977.

Francis Harry Compton Crick was born in 1916 and educated at Mill Hill School and University College, London. His first degree was in physics. During the second world war he was a scientific civil servant in the Admiralty, working on the design of such devices as magnetic and acoustic mines.

After the war he turned to biological research, going first to the Strangeways Laboratory, Cambridge, on a Medical Research Council Studentship and, in 1949, joining the Medical Research Council Unit in the Cavendish Laboratory at Cambridge as a Laboratory Scientist until 1961. He obtained his Ph.D. from Cambridge in 1954.

Dr. Crick's research originally concentrated on x-ray diffraction studies of proteins. With three other scientists (Professor Maurice Wilkins and the late Dr. Rosalind Franklin at King's College, London, together with Dr. James Watson), he was responsible in 1953 for the discovery of the molecular structure of deoxyribonucleic acid (DNA)—the biological structure which makes possible the transmission of inherited characteristics. Subsequently he worked on problems connected with protein synthesis and the genetic code. After 1966 his interest turned to developmental biology and chromatin structure. Since joining The Salk Institute in 1976, Dr. Crick's research work has been mainly in neurobiology, and especially in the visual system of mammals. He has also written on the neural basis of attention, REM sleep, and on visual awareness.

Dr. Crick was elected a Fellow of the Royal Society in 1959. He is a foreign member of several national academies including the U.S. National Academy of Sciences. In 1960 he was a recipient of an Albert Lasker Award of the American Public Health Association for work in medical research. The following year he was awarded the Prix Charles Leopold Mayer of the French Academy of Sciences and the Research Corporation Award. In 1962 he received the Gairdner Award of Merit for outstanding medical research from the Gairdner Foundation in Toronto, to be followed by the Nobel Prize for Physiology and Medicine, shared with Professors Watson and Wilkins.

He married Odile Speed in 1949; they have two daughters. Dr. Crick has a son by an earlier marriage.

George Porter, 1986.

Lord George Porter

Lord Porter is Chairman of the Centre for Photomolecular Sciences at Imperial College, Biology Department, London SW7 2BB; Chancellor of the University of Leicester, Professor of Astronomy and Other Physical Sciences, Gresham College, and Emeritus Professor of Chemistry of the Royal Institution of Great Britain.

He was born in 1920 and educated at Thorne Grammar School, Leeds University and Emmanual College, Cambridge, where he is an honorary fellow. Between 1941 and 1945 he served as a Radar Officer in the Royal Navy. On completion of his doctorate degree he spent a further five years as Demonstrator and then Assistant Director of Research in the Physical Chemistry Department of the University of Cambridge. He was appointed Professor of Physical Chemistry in the University of Sheffield in 1955 and head of the Chemistry Department in 1962. In 1960 he was elected a Fellow of the Royal Society and he shared the Nobel Prize for Chemistry in 1967. He was Director of the Royal Institution of Great Britain from 1965 to 1985. He was President of the Chemical Society of London from 1970 to 1972; the National Association for Gifted Children 1975/80, the Association for Science Education in 1985, and the British Association for the Advancement of Science 1985/86. He was President of the Royal Society from 1985 to 1990. He served as a trustee of the British Museum, on the Science Research Council, and is a member of the Government's Advisory Council on Science and Technology.

He was knighted in 1971 and elevated to the peerage as Lord Porter of Luddenham in 1990. He has lectured widely at home and abroad and has received many honours and awards in recognition of his work, including the Order of Merit, the Davy and Rumford Medals of the Royal Society, and the Kalinga Prize for the Popularisation of Science. He holds honorary degrees of 30 universities, seven honorary professorships and fellowships, and is an honorary member of many foreign academies.

His research interests are in the field of fast reactions, photochemistry, photosynthesis and solar energy. He is also interested in scientific education and the presentation of science to nonspecialists and has taken part in numerous BBC television programs, including 10 lectures on "The Laws of Disorder" and two series of Royal Institution Christmas Lectures on "Time Machines" and "The Natural History of a Sunbeam."

Photo of Dudley Herschbach taken by Bob Campagna April, 1988 at Cornell College, Iowa.

Dudley R. Herschbach

Dudley Robert Herschbach was born in San Jose, California (1932) and received his B.S. degree in mathematics (1954) and M.S. in chemistry (1955) at Stanford University, followed by an A.M. degree in physics (1956) and Ph.D. in chemical physics (1958) at Harvard. After a term as Junior Fellow in the Society of Fellows at Harvard (1957–1959), he was a member of the Chemistry Faculty at the University of California, Berkeley (1959–1963), before returning to Harvard as Professor of Chemistry (1963), where he is now Baird Professor of Science (since 1976). He has served as Chairman of the Chemical Physics program (1964–1977) and the Chemistry Department (1977–1980), as a member of the Faculty Council (1980–1983), and Co-Master with his wife Georgene of Currier House (1981–1986). His teaching includes graduate courses in quantum mechanics, chemical kinetics, molecular spectroscopy, and collision theory, as well as undergraduate courses in physical chemistry and general chemistry for freshmen, his most challenging assignment.

He is a Fellow of the American Academy of Arts and Sciences, the National Academy of Sciences, the American Philosophical Society, and the Royal Chemical Society of Great Britain. His awards include the Pure Chemistry Prize of the American Chemical Society (1965), the Linus Pauling Medal (1978), the Michael Polanyi Medal (1981), the Irving Langmuir Prize of the American Physical Society (1983), and the National Medal of Science (1991); he shared with Yuan T. Lee and John C. Polanyi the Nobel Prize in Chemistry (1986).

Professor Herschbach has published over 250 research papers. His current research is devoted to molecular beam studies of reaction stereodynamics, intermolecular forces in liquids, and a dimensional scaling approach to electronic structure.

John Polanyi being interviewed in his hotel room in Montreal.
Photo by Mr. Dave Sidaway, of 'The Montreal Gazette', September 3, 1988.

John C. Polanyi

John Charles Polanyi, educated Manchester University, England, and Princeton University, U.S.A., joined the University of Toronto, Canada, in 1956. His research is on the molecular basis for chemical reaction. He is a Fellow of the Royal Societies of Canada, of London, and of Edinburgh, also of the American Academy of Arts and Sciences, the U.S. National Academy of Sciences and the Pontifical Academy of Rome. His awards include the 1986 Nobel Prize in Chemistry and the Royal Medal of the Royal Society of London. He served on the Prime Minister of Canada's Advisory Board on Science and Technology and presently is Honorary Advisor to the Max Planck Institute for Quantum Optics, Germany, the Steacie Institute for Molecular Sciences, Canada, and the Institute for Molecular Sciences, Japan. He was the founding Chairman of the Canadian Pugwash Group in 1960, has written extensively on science policy and on the control of armaments, and is co-editor of a book, *The Dangers of Nuclear War.*

Ahmed Zewail, with Linus Pauling, on the Caltech campus, July 17, 1990.

Ahmed Zewail

Ahmed H. Zewail was born in Egypt in 1946, where he received his early formal education at Alexandria University (B.S. 1967 and M.S. 1969). He went to Philadelphia, Pennsylvania, in 1969 and after earning his Ph.D. from the University of Pennsylvania in 1974, he joined the University of California at Berkeley as an IBM Research Fellow. Shortly afterward (1976), he joined the faculty of the California Institute of Technology. In 1990, he was honored with the first Linus Pauling Professorship of Chemical Physics at Caltech.

Professor Zewail's honors and awards include the King Faisal International Prize in Science, the Alfred P. Sloan Fellowship, the Camille and Henry Dreyfus Foundation Teacher-Scholar Award, the ACS Harrison-Howe Award, Guggenheim Foundation Fellowship, the Alexander von Humboldt Award, the Hoechst prize, the Carl Zeiss Award, and the Nobel Laureate Signature Award for graduate education. He has held visiting professorships in the US, Holland, France, Germany and UK. Professor Zewail is a member of the National Academy of Sciences, the Third World Academy of Sciences, and a fellow of the American Physical Society.

Professor Zewail is the author and co-author of more than 200 scientific papers, and the editor of four books on laser chemistry and spectroscopy, in addition to *The Chemical Bond: Structure and Dynamics*. He is also the U.S. editor of *Chemical Physics Letters*. His current research interests are in the field of ultrafast laser chemistry–femtochemistry. He is married to Dr. Dema Faham, and proud of two daughters (Maha and Amani); one of them, Maha, is an undergraduate student at Caltech.

Richard Bernstein (second from the left) in the last photo taken of him with colleagues in Leningrad (now St. Petersburg) on June 20, 1990. Photo courtesy of the Wittig family.

Richard Bernstein

Richard B. Bernstein, born in New York City on the thirty-first of October, 1923, was educated in the public schools of New Jersey. He received an A.B. with honors in chemistry and mathematics in 1943, and an M.A. in chemistry in 1944, both from Columbia University in New York. As a researcher on the Manhattan Project at Columbia from 1942 to 1946, Professor Bernstein worked on the uranium isotope separation program. During part of that period, from 1944 to 1946, he served in the United States Army Corps of Engineers, Manhattan District. In a military capacity, he took part in the Bikini operations in the Pacific, returning to Columbia University shortly afterward to complete his Ph.D. (1948).

Professor Bernstein's research interests were in the area of physical chemistry and chemical physics, including work in femtochemistry (with Professor Zewail), isotope separation, reaction kinetics and catalysis, vibrational and electronic spectroscopy, inter-molecular forces, collision phenomena, molecular beam scattering, laser chemistry, and, especially, the dynamics of elementary chemical reactions.

Professor Bernstein was a professor of chemistry at the University of California, Los Angeles, from 1983 until he passed away. He had previously been the Higgins Professor at Columbia University and had also taught chemistry at a number of universities, including the Illinois Institute of Technology, the University of Michigan, of Wisconsin, and of Texas. He had also served as the Senior Vice President of Occidental Research Corporation.

Among the numerous awards Professor Bernstein gathered during his lifetime were the ACS Peter Debye and Langmuir awards, the Chemical Sciences Award of the National Academy of Sciences, the 1988 Welch Award in Chemistry, and the National Medal of Science in 1989.

Professor Bernstein died in Helsinki on July 8, 1990, survived by his wife, Norma, and four children (Neil, Minda, Beth, and Julie).

Group photo, courtesy of the California Institute of Technology,
taken on February 28, 1991.

Group Photo

Left to Right:

Fred Anson
Alexander Rich
Dudley Herschbach
John Polanyi
Tom Everhart
Linus Pauling
Max Perutz
John Roberts
George Porter
Ahmed Zewail
Norman Davidson
(Francis Crick arrived after the photo was taken—please see
page 94.)

INDEX